Igniting the Caribbean's Past

Igniting the Caribbean's Past

Fire in British West Indian History

Bonham C. Richardson

The University of North Carolina Press
Chapel Hill and London

Designed by Gary Gore
Set in Cochin
by Keystone Typesetting, Inc.
Manufactured in the United States of America
The paper in this book meets the guidelines for permanence and durability of the Committee on Production Guidelines for Book Longevity of the Council on Library Resources.

Library of Congress Cataloging-in-Publication Data
Richardson, Bonham C., 1939–
Igniting the Caribbean's past : fire in British West Indian history / Bonham C. Richardson.
p. cm.
Includes bibliographical references and index.
ISBN 0-8078-2854-8 (cloth: alk. paper)
ISBN 0-8078-5523-5 (pbk.: alk. paper)
1. West Indies, British—History—19th century. 2. Fires—West Indies, British—History—19th century. 3. Arson—West Indies, British—History—19th century. 4. Slave insurrections—West Indies, British—History—19th century. 5. Plantation life—West Indies, British—History—19th century. 6. West Indies, British—Race relations. I. Title.
F2131.R53 2004
972.9′7—dc22 2003019679

cloth 08 07 06 05 04 5 4 3 2 1
paper 08 07 06 05 04 5 4 3 2 1

In Memory of Barney Nietschmann, 1941–2000

Contents

Illustrations

Preface

Caribbean history is punctuated by recurring earthquakes and hurricanes, natural hazards that have played momentous roles in the unfolding of events in a region also influenced by insularity, colonization, and gunboat diplomacy. Fire also has influenced the region's history and geography in important ways. Fire has cleared forests, burned sugarcane, sparked slave insurrections, attracted crowds, lighted streets and houses, and symbolized protest in the region for centuries. Human beings, not natural forces, have ignited nearly all Caribbean fire. So the personality of fire in the region is closely associated with the human history of the Caribbean, so much so that writers and poets often use "fire" to signify their discontent with the Caribbean's steeply tiered social hierarchy, which is itself an extension of the region's past.

This book is a historical geography of the Lesser Antilles (the island arc at the far eastern end of the Caribbean) from about 1885 to 1910. It is a study of how the common people of the region—inhabiting a string of tiny, hazard-prone, depression-racked, economically and environmentally worn-out British island colonies—used, modified, and contemplated fire. Fire played a variety of daily and seasonal roles there, and as in all human societies, the people of these small islands resorted to the use of fire at different times, for many reasons, and on different scales, in both utilitarian and symbolic ways in living out and attempting to improve their individual and collective existences.

Because this study is set in the past, it will be of interest to students of history. Yet its approach is geographical; this means, without embarking on a tedious discussion of what academic geography is or should be, that it attempts to be holistic, trying to be at once social, environmental, political, and in this particular case, colonial. Especially at the microscale, where this study takes place, efforts to disaggregate these intertwined themes or topics risk treating academic categories as things and as the bases for extracting ideas and people and material objects from their mundane contexts, thereby falsifying reality and reducing understanding. A holistic view is by no means an attempt to be all-encompassing; the focus or integrating theme here is the human use of fire, which is intended to knit together the various topics (social, environmental, and so forth). Accordingly, and as discussed in the pages that follow, Caribbean fire at the end of the nineteenth century and into the early twentieth went far beyond the combustion of flammable materials. Among many other things, fire could at once awaken the past, create opportunity or fear, and, literally, illuminate and thereby throw into high relief events occurring in the Lesser Antilles more than one century ago.

I began concentrating on this era of British Caribbean history in the late 1960s without knowing I was doing so. Early in my academic career my interests lay in local-level livelihood activities in Guyana and then Trinidad. In every village I visited in these two countries, there were old people whose recollections extended back to the turn of the century and even earlier. I sought them out for "background" information and learned about early estate labor regimens, the coming of the first automobiles, the contrasts—as they saw and experienced and remembered them—between cities and countrysides, and innumerable other topics. As my field research moved north into the Lesser Antilles, to Grenada, St. Kitts–Nevis, and Barbados, in the 1970s and early 1980s, I began to concentrate more and more on these local historical issues, extending and augmenting my earlier interviewing with conventional archival sources. All of the old informants I interviewed in these places have died, but I have continued to follow the leads and ideas and perspectives they first provided for me into the archives, keeping in mind the written and oral recollections they were so generous to share with me.

The initial impetus for this book on Caribbean fire came from a conversation with Virginia Tech geography department colleagues perhaps a decade ago. As we compared our field experiences in different parts of the world, I found myself suggesting that, unique among the events I had witnessed in the small islands of the Caribbean, a sugarcane fire had the virtue of seeming to pull everything together, being at once an environmental, social, and biological event and therefore an eminently geographical phenomenon. It occurred to me then that Caribbean fire might be a promising book topic, so I began to note and file references pertaining to fire, although I had other more immediate and pressing writing obligations at the time. Little did I know then, as I think I know now, how widespread and diffuse the written historical information about Caribbean fire can be. As scholars of fire know, their subject is everywhere and nowhere. The same might be said about archival information about fire. Similarly, all Caribbean specialists would certainly agree that fire plays many roles in the region's history, and one can hardly conceive of descriptions of, as examples, a protest or slave insurrection without a backdrop of flames. Yet it is nearly impossible to find "fire" in the indexes of the books academics write about these and other events.

The scattered and elusive nature of archival evidence of fire has meant the turning of many pages and, more often, the scanning and searching of innumerable screens of microfilm, and I am most grateful to the several agencies that funded the travel and time that went into the preparation for and writing of this book. The National Geographic Society, almost surely against the better judgment of some there who decry exclusively archival research, provided a valuable subsistence and travel grant for work in London libraries; I hope that this book's appearance will compensate. My wife, Linda Richardson (a reference librarian as well as my travel partner, research companion, and best friend for thirty-five years), and I also received a research grant from the Anne U. White Fund of the Association of American Geographers, a program intended to encourage the joint work of association members and their spouses. The Virginia Tech Humanities Summer Stipend Program supported my research and writing for this project on three separate occasions, and I am particularly grateful for the program's confidence and patience in this and others' long-term academic projects. A Virginia Tech research assignment for the fall of 1998 allowed me a semester during which I was relieved of teaching duties so that I could visit other libraries, but mostly I reeled and read microfilm in my office.

My wife, our younger daughter, and I rented flats in London during the summers of 1996, 1997, and 1998, when my wife and I worked in various libraries there. Valuable information about British Caribbean history, including special reports about particular fires, fire legislation, and fire brigades, was available at the Public Record Office and in the holdings of the Institute of Commonwealth Studies at the University of London and at the Foreign and Commonwealth Office archives. The incomparable library at the Herbarium and Archives at the Royal Botanic Gardens at Kew, with its holdings of official correspondence, a variety of early agricultural reports, and the British Indian forester E. D. M. Hooper's published surveys from his sojourn in the Lesser Antilles in the 1880s, provided key information about environmental issues.

The most valuable published information, representing the major source of evidence on which this study is based, came from the early West Indian newspapers at the British Library's newspaper archives at Colindale, in the northern part of greater London, as well as the microfilmed newspapers that I obtained via interlibrary loan from Tulane University and the University of Florida collections. In a very few cases I have cited newspaper evidence from earlier research periods in the Caribbean, specifically from St. Kitts in 1976 and Barbados in 1980 and 1981–82.[1]

Because these early newspapers served special interests when they were published, they are shot through with what some might condemn as elitist structured bias. Yet they often provided workers' perspectives as well. The newspapers' most obvious virtue is their vivid, immediate descriptions of events on the ground from day to day. And they provide valuable local context unavailable in other archival sources. In proceeding from January to December, to take only one such example, one is impressed with the seasonality of local Caribbean fires, their high incidence in the dry early months of the year in direct contrast with their near absence in the wet months of July to November. Nearly as important, West Indian papers at the turn of the century also told of contemporaneous events elsewhere, in effect presenting a global canvas on which to place events and issues occurring on individual Caribbean islands and displaying the news that competed for the attention of, as well as influencing, local decision makers. Headlines bemoaned the catastrophic eruption of Mount Pelée in Martinique in May 1902, which had obvious implications for the whole region. The Barbados newspapers featured battle-by-battle details of the Boer War (1899–1902) in South Africa, reflecting the relatively large white population on the island and the recruitment of young white Barbadian men into the British army. Occasional notices in several papers reminding "readers at Colón" to pay for their subscriptions underlined Panama's importance as a destination for human migrants and indicated that news published locally traveled far.

Yet the great importance of early British West Indian newspapers was that they both mirrored and molded events at the time. Early in 1901, for instance, the *Barbados Agricultural Reporter* lamented that lawlessness on the island was heightened because the average worker "reads in the newspaper accounts of strikes, lock-outs, and other forms of conflict between capitalists and labourers in big countries in Europe and America, and loses no time in jumping at the conclusion that his . . . condition is . . . desperate . . . indeed."[2] It is likely that local officials in the British Caribbean at the time, not to mention historically oriented academics a century later, did not fully appreciate that such news in local papers was subsequently disseminated widely and probably embellished by word of mouth, helping to fuel and intensify the local resentment of prevailing inequities.[3]

As in any project of this kind, I received a great deal of help, kindness, encouragement, and advice from others in pursuing my work. In London, Gad and Ruth Heuman helped us find suitable summer rental units close to public transportation. Librarian Lesley Price at the Royal Botanic Gardens provided particularly expert guidance and assistance. The hospitality of geographers David and Mary Alice Lowenthal was, as usual, as academically stimulating as it was pleasant socially, and on a number of occasions at their home over dinner I had the opportunity to raise points or entertain issues I had encountered that same day in the archives.

Colleagues from the Caribbean also offered encouragement and helpful critiques as the writing progressed. Since the preparation of this book extended over several years, I was able to offer parts of what eventually became book chapters as conference papers. In April 2000 I presented a draft of what would become part of Chapter 4 at the annual meetings of the Association of Caribbean Historians in French Guiana. In April 2002 I gave a paper about early sugarcane fires—which would become part of what is now Chapter 5—at the Association of Caribbean Historians conference in the Bahamas. I am indebted to Selwyn Carrington, Mary Chamberlain, Woodville Marshall, and others for both encouragement and critique at those meetings.

Others also deserve special thanks. Librarian Sue Fritz at Virginia Tech went well beyond the call of duty in providing access to microfilms and film readers. My departmental colleague and fellow Caribbeanist Larry Grossman is an unfailing source of editorial wisdom, common sense, and good humor. Jerry Handler, from the University of Virginia, provided encouragement, an opportunity to present early ideas in a seminar there, and tough questions. Deborah De, Sara Beth Keough, and Anna Ward gave me generous and much-needed computer help at crucial junctures. Bridget Brereton, Jim Campbell, Jim Craig, John Davies, Harm de Blij, Bill Denevan, Alan Dye, Sue Farquhar, Jock Galloway, Lisa Kennedy, Richard Price, and Stephen Pyne dropped what they were doing at various times to lend support, answer specific questions, or point out relevant references before and during the writing. Jean Besson and Karl Watson graciously provided photo permissions. Elaine Maisner at the University of North Carolina Press was most encouraging and helpful, and she put the book manuscript draft into the hands of two anonymous, knowledgeable, and careful readers who forced me to think harder about a clearer presentation. Stephanie Wenzel's expert copyediting improved the book substantially. Richard and Laurie Shepherd generously offered the use of their country home near Blacksburg, Virginia, as a place for our family to stay in August 2003, and as a quiet haven for reviewing the copyedited manuscript.

This book is dedicated to the memory of an eminent geographer and close friend. Bernard Nietschmann died of esophageal cancer in January 2000 at age fifty-eight.[4] His memorial gathering at Berkeley, California, in May of that year attracted participants from the United States and Canada, Europe, Central America, and Mexico. He and I were friends as graduate students at Wisconsin, and we finished the Ph.D. degrees one day apart. Yet Barney set the bar so high that none of the rest of us ever could really reach it. Our intellectual leader at Madison, he subsequently became a renowned teacher and scholar at Ann Arbor and then Berkeley and, more important, a fierce advocate of the rights and resources of the Miskito Indians of coastal Nicaragua. His advocacy for the well-being of indigenous peoples took him around the world, winning him acclaim, which he never sought, and a few political enemies, with whom he found little need to compromise. Barney was particularly encouraging when I began this book project, and a year or so before his death he traveled to Barbados and St. Lucia seeking interest and support for the kinds of environmental protection programs that he sponsored so successfully in the far western Caribbean.

Igniting the Caribbean's Past

1

Introduction

Whoever feared earthquakes, would erect a house of wood.
Whoever feared hurricanes or remembered the fire,
erected a house of stone.

Patrick Chamoiseau, *Texaco*

THE SUGARCANE PLANTATION OWNERS ON THE British island colony of Antigua, in the northeastern corner of the Caribbean Sea, considered themselves under siege. For a decade their crops had brought markedly lower prices than in 1884, the year that, through no fault of their own, the price of raw sugar plummeted in London. The United States, with its own protected market, not only offered no outlet for Antigua's sugar but also was financing competing cane sugar production in the Greater Antilles. Now, in late March 1895, these traumatic external conditions were intensifying the immediate, day-to-day agricultural conditions on Antigua itself. The low-lying island's characteristically dry weather had reduced crop outputs even more than usual. Far worse, a few of the disgruntled and desperately impoverished black laborers had begun to resort to fire, one of the planters' worst nightmares. By secretly burning fields of growing canes, the workers were at once performing acts of vengeance and attempting to force emergency harvesting work to provide badly needed wages.

Although the identities of the originators of the cane fires were, as usual, secrets well hidden from the local officials and planters, the windswept blazes themselves reinforced and publicized Antigua's socioeconomic malaise because of the fires' contagious social effects. The flames and smoke from a single large cane fire towered above the island's arid landscape, bringing firefighters as well as the horses, rolling stock, and associated clatter of the fire brigade and thereby attracting energized crowds of onlookers. On the day or night of a major sugarcane fire on a small island like Antigua, nothing else was worth talking about. Local newspapers condemned the fires. Planters fumed in anger and frustration. Government officials dutifully sent memos to London detailing property losses and explaining how the economy was threatened. Insurance agents warned of higher premiums. It takes little imagination to suggest that conversations inside the small wooden houses occupied by the black laborers' families on such a day probably were dominated by exclamations about fire, the columns of smoke, and the excitement of it all. Furthermore, some of these conversations doubtless expressed sympathy with the incendiaries and must have suggested more burning.

These were understandable attitudes given the repressive social atmosphere of the British Caribbean in the late 1800s, where black working classes only two generations removed from slavery were ruled by a tiny minority of white planters and a cadre of local white officials representing the London Colonial Office. The planters' and officials' attitudes about the cane fires and much else usually could be found in the island's weekly newspaper, the *Antigua Standard*. In condemning the series of covert sugarcane fires early in 1895, the newspaper warned that the blazes posed problems "of almost unprecedented severity" and spoke out against these "alarming" events perpetrated by "cowards who . . . indulge their passions."[1]

But the burning and its consequences were by no means confined to Antigua. Incendiarism flared on the neighboring islands, too, and taken altogether, the blazes constituted a regional sugarcane fire crisis. Low sugar prices had created similar depression conditions throughout the British Caribbean, and beleaguered laborers on several of the island colonies had responded by torching the growing cane fields, just as they had in Antigua. In June 1898 an alarming letter appeared in all of the newspapers in the region from the office of the London law firm of Malcolm Kearton & Company Ltd., which represented commercial firms throughout the British Caribbean. It warned of prohibitive rates that London insurance companies soon would have to charge to insure the growing crops; the letter further suggested that the fires might cause businesses even to withdraw capital from the region, which already was woefully short of financial backing. The notice ended by expressing "grave concern" about the safety of investing in the British Caribbean unless local authorities proceeded with "the greatest rigour" to apprehend those responsible for the illicit fires.[2]

Cane fires were only one among several burning issues that distressed local island officials in the British Caribbean at the turn of the twentieth century. The depression also helped to trigger arson in towns and cities. Rural dwellers routinely set "bush fires" in the countrysides that posed dangers to forests and adjacent settlements, especially during droughts and dry seasons. The ominous torchlight rituals of the *cannes brulées* (burning canes) processions in Trinidad and elsewhere pitted black urban dwellers against local police—confrontations often marred by violence—at the beginning of every Lenten season. Flames in general, especially those transported from place to place by crowds carrying torches, indicated discontent and possible disturbances, deep-seated social signals rooted in slavery. And the aging, sun-bleached, wooden buildings at the centers of the region's capital towns and cities were frighteningly vulnerable to accidental fires. Especially in the tense social atmosphere underlain by economic depression, the human use of fire in the fin-de-siècle British Caribbean was as great a local and regional threat as an earthquake or a hurricane.

This book is a historical geography of the small British island colonies of the eastern Caribbean in the depression decades of the late nineteenth and early twentieth centuries; its subject is the people of the region and their interrelationships with fire—fire as it was used, categorized, contemplated, contested, and feared by West Indians in both rural and urban areas at the time. Similar to other studies in historical or cultural geography, this study dwells on mundane, grassroots, environmental phenomena as they have influenced and are affected by wider issues and events.[3] Caribbean fires at the turn of the century, as exciting and important as they were in their local manifestations, could not be fully understood outside the overall context of economic depression that had its origins in British and European trade. And the repercussions of the local fires, as exemplified by the 1898 insurance letter from London, reached well beyond the places of their occurrence.

Fire is the principal theme of this book; it is not here intended as the overriding causal agent or single variable that best explains the human geography of the British Caribbean at the turn of the century. Accordingly, relevant digressions appear in the pages that follow—from discussions about the origins of tropical forestry as a science to speculations about the importance of the Caribbean's human migration patterns—intended to augment a fuller understanding of fire's role in British Caribbean history. The importance of fire in this study, in other words, is not intended to identify the One Big Cause or One Big Story in the environmental history of the eastern Caribbean.[4] On the other hand, one would be hard pressed to convince the residents of Kingston, Jamaica, in 1882 or Fort-de-France, Martinique, in 1890 or several other large cities in the Caribbean region (not to mention the sugar plantation owners in Antigua in 1895) of these theoretical fine points after their cities or estates had been burned to the ground. Nor were the damaging rural fires and enormous city fires of the era important simply because of the immediate devastation and misery they caused. The huge fire in Port of Spain, Trinidad, in early March 1895, for example—an event that historians of the Caribbean have scarcely noticed—directly influenced international negotiations, political wrangling that reached as far as London, local and regional social transformations, and Trinidad's economic infrastructure.

Caribbean history in general, furthermore, would have unfolded very differently without the ways in which people have used fire. The nature, timing, and extent of, among other things, environmental transformations, agricultural regimens, and sociopolitical demonstrations in the region all have depended heavily at some time on the human use of fire. While it may be fairly argued that fire has been a catalyst rather than a determinant in Caribbean history and geography, the same might be suggested about sugarcane, insularity, and hurricanes. In the end, the principal virtue of studying fire in Caribbean history is that it emphasizes and underscores the material circumstances of lives lived there in the past. Accordingly, I hope to throw light, using fire as a window or point of entry, on understanding better what things were like on the ground in the islands one century ago. Different from written descriptions of many other events at the time, accounts of fire—like the fires themselves—usually were spontaneous, immediate, and not always masticated and digested (and thereby distorted) by official interpretation. Burning buildings or blazing sugarcane fields attracted onlookers from all segments of local societies, inspired immediate preventive measures, invited speculations as to origins, and led to on-the-spot reporting that enlivened the texts of agricultural reports and newspapers on these small islands. Reading and researching the accounts and reports about these turn-of-the-century fires in the Lesser Antilles, one might argue, is about as close as we can get to being there.

A historically informed assessment of human-fire interrelationships in the Caribbean of 100 years ago represents an appealingly specific and fresh perspective different from the approach and subject matter normally found in the conventional socioeconomic histories of the region based on political correspondence and officially published government commentary. The latter, while revealing and important, invariably dwell on the actions and thoughts of decision makers in London and their bureaucratic counterparts in Bridgetown, Barbados, or Port of Spain, Trinidad, or the capital towns of the other islands of the region. Although many of the studies based exclusively on this information are richly researched and imaginatively interpreted, they offer only indirect insight into the lives of the vast majority of the people who inhabited the island colonies and whose activities shaped and directed the local histories of these places. The approach taken here—although based on written records—is intended to exemplify or to fit in with the recent academic thinking that tries to remedy or offset the structuredness of colonial history and the partial truths often contained in conventional archival records.

Fires and similar environmental events are and were vivid, immediate, and perhaps most important, all-encompassing because they necessarily involved and affected and influenced everyone—not just the governors, military commanders, and large-scale planters of the island colonies who called meetings and left written records. Grassroots geographical events, furthermore, often had long-term implications for individuals and families, influencing, for example, decisions to emigrate or to build new houses or to participate as members of torch-bearing crowds or similar groups whose actions influenced local changes. A specific example concerning published Caribbean scholarship perhaps clarifies the need for inclusive, ecological approaches to the region's social history: A recent sociopolitical history of Jamaica, covering much of the nineteenth and the early twentieth centuries, has been appropriately lauded for its simultaneous attention to events in Britain and Jamaica and its resulting wide perspective. Yet this particular study mentions neither the massive fire that destroyed much of central Kingston in December 1882 nor the catastrophic earthquake that leveled the central part of the city in January 1907, two events that represented perhaps the most profound and memorable experiences for an entire generation of Jamaicans who lived there during the time period covered by the book.[5]

Except for a brief regional overview of fire since 1492 in the following chapter, my study does not attempt to capture the subject of Caribbean fire in its entirety. Rather, I delimit Caribbean fire in time and space. I also suggest that fire has a particular personality or meaning in the Caribbean not always found in other regions of the world.[6] The Caribbean's insular and fragmented physical environments have been severely modified and degraded by human occupation for several millennia; such modification—often by fire—has been particularly notable in the five centuries since European intrusion. Therefore fire in the Caribbean does not connote the potential for vast and catastrophic environmental danger to a pristine wilderness as it does in, say, the high-latitude forests of Eurasia and North America and, more recently, the tropical forests of Southeast Asia.[7] Nor is fire in the Caribbean so closely interrelated with the evolution of the region's flora and fauna that it is viewed as it is on the Australian "fire continent."[8] Rather, Caribbean fire carries with it a strong social connotation, a distinction created by the region's human history of remarkable and persistent social oppression exerted by local planters and other officials, oppression that has in turn been met by the resistance and creativity of the region's working peoples. Slavery, insurrections, threats, and planned destruction in the region all have been accented by either fire or threats of fire.[9]

The Caribbean's unique colonial history also helps to explain why ecologically oriented approaches to understanding the region's past are relatively uncommon. The Caribbean was Europe's first transatlantic colony, its aboriginal inhabitants swept away by enslavement and disease in only a few decades. Then a "native" population from elsewhere was subsequently introduced, first as slaves and later as indentured laborers, by European planters who invariably interposed their demands for forced labor between the introduced working peoples and the land itself. This is why there are few scholarly "man-land" studies of Caribbean peoples that assume essential relationships between rural cultivators and the lands they inhabit, as there are, for example, in mainland Latin America. The Caribbean's unique environmental history also helps to explain why the relatively recent academic approach known as "political ecology," whose practitioners insist that famines, droughts, and other "environmental" events and phenomena are more often man-made than they are God-given, seem second nature to most Caribbeanists.[10] The fundamental and historically produced separations between people and the land they occupy in the Caribbean region also help us understand why the men and women of the region have migrated so readily once they have been free to do so, as well as why they have been so prone to resist and protest and demonstrate, actions animated by, among other things, the use of fire.

The Study Region and Time Period

Although it focuses on the human use of fire, this study also intends to provide a social geographical portrayal of the British-owned Lesser Antilles at the end of the nineteenth century and the start of the twentieth. Extending from Puerto Rico and the Danish Virgin Islands (which would be sold to the United States in 1917) south to Venezuela, this insular arc of British possessions was punctuated by three Dutch islets in the north and the larger French territories of Guadeloupe (and its outliers) and Martinique farther south. Taken altogether, the British territories in the eastern Caribbean at the turn of the century included the British Virgin Islands in the north, the British Leewards farther east, and the British Windwards and Barbados to the south, eventually encompassing Trinidad and Tobago, just north of Venezuela (see map). Whereas all of these islands shared important social and geographical characteristics—political subservience to the British Colonial Office, a black Afro-Caribbean or mixed-blood working class dominated by a small number of whites, an economy based on the production of tropical cash crops (usually sugarcane) for export, and the deceptively simple fact of insularity—each place was very different from the others, a point far more obvious and meaningful to the people who lived there than it was to the metropolitan officials sent from Britain and their superiors based in London. Nor is the geographical delimitation of this study area intended to ignore the importance of nearby areas such as the French Antilles or Central America at the time because these neighboring areas, as well as others outside the region—notably the United Kingdom and the United States—played important, if usually indirect, roles in the lives of the people of the eastern Caribbean at the turn of the twentieth century.

Because small islands are far more vulnerable to natural or human-induced shock or calamity, including fire, than are large islands, the very small sizes of the Lesser Antilles were and are more meaningful geographically than their simply being recognized as places with diminutive land areas. The famous ecological monograph by Robert H. MacArthur and Edward O. Wilson dealing with island biogeography pointed out years ago the correlation between Caribbean island size and number of native faunal species.[11] Subsequent regionwide research focused on the Caribbean's natural vegetation has shown the same to be true with floral species.[12] The ecological rule of thumb that the greater the number of species, the greater an ecosystem's complexity and associated resistance to natural or man-made damage or conversion could not be more relevant to a historical study of the role of fire in small places. In such places, large fires (or passing hurricanes or earthquakes or drought conditions) affect severely not just districts but entire islands, and such events have occurred more than once in the history of the Lesser Antilles.

Small-scale insularity carries with it important economic and cultural attributes as well. A tiny island specializing in one crop or even one crop species or a single economic pursuit experiences greater economic volatility than do larger places, not to mention greater vulnerability to calamities such as heavy rains, pest infestations, or the severing of telegraph lines. Short distances between village communities or between villages and capital towns facilitate the rapid transfer of material goods and information. People living in small islands grow accustomed to seeing a limited number of familiar faces, and strangers, in contrast, stand out. A finite number of potential conjugal partners often means that many people in one small island are related by blood. So it is not uncommon in the Lesser Antilles for a high percentage of the human inhabitants to possess the same surname, a situation that inspires novel ways of distinguishing one individual from another.[13]

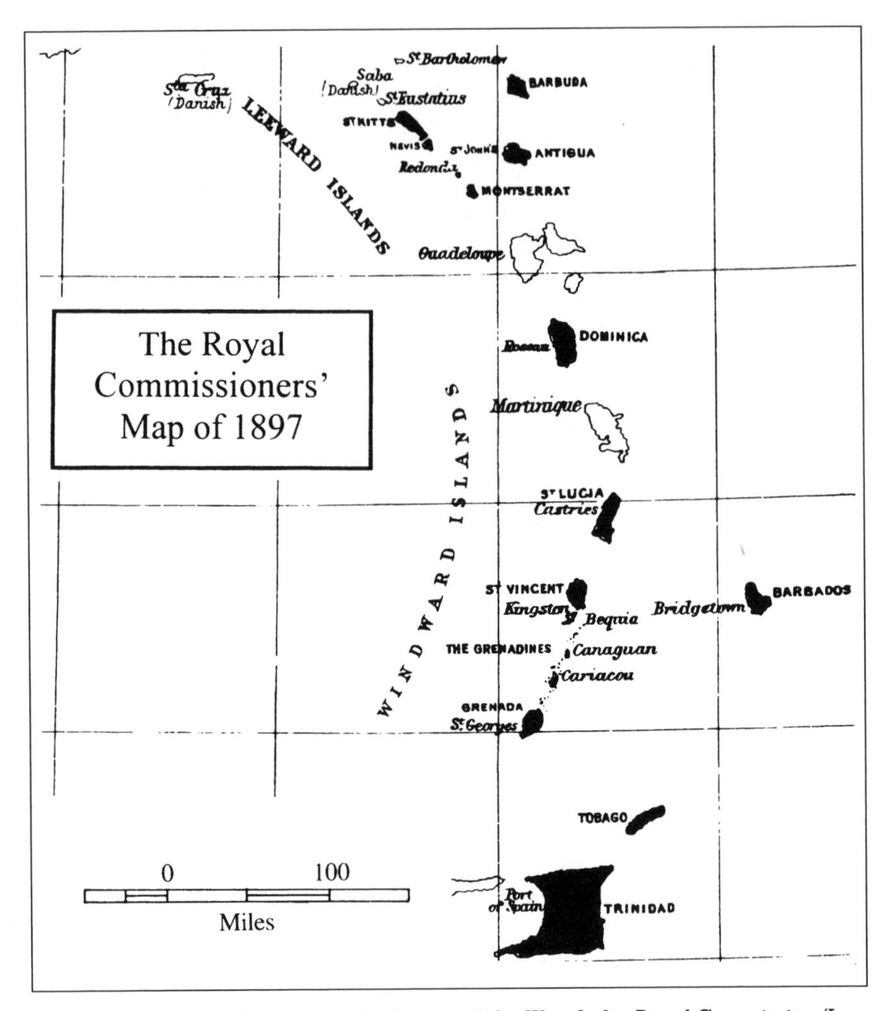

A portion of the map illustrating the *Report of the West India Royal Commission* (London: HMSO, 1897)

When an extraordinary event occurs on a very small island, such as a political demonstration, the arrival of a notable ship passenger, or the occurrence of a particularly large fire that is visible from practically every point on the island (such as the sugarcane fires in Antigua early in 1895), the whole place can become enlivened or vitalized.

Whereas the small size and particularity of each island matter greatly, it is equally important to underscore the regional view taken in this book. A study of the small British islands of the eastern Caribbean in their entirety is unlike that of a single island or even of politically paired islands such as

Trinidad and Tobago or Antigua and Barbuda. A regional view loses the detail that concentration on a single place provides, and a regional outlook inevitably emphasizes some places to the detriment of others. Yet such an approach carries with it the virtue of portraying the broad and fluid regional habitation spaces in which real people lived, not the spatially delimited single-island colonies that proved so convenient for colonial record keepers, convenience that likely obscures a sufficiently wide view for subsequent researchers relying on the information that the record keepers left behind. To be sure, most West Indians in the late 1800s lived their entire lives on a single island. Yet they usually could see, and often talked with one another about, the neighboring islands and had friends and relatives who lived there.

The relatively underresearched time period covered by this book might appropriately be considered a temporal bridge between the present and the past. The great-grandparents of today's adult West Indians were living in the islands then. It also was an era when a few very old people there had personal recollections of plantation slavery, remembrances they regularly shared with friends and family. So in 1888, for example, when the proposed torchlight parades to celebrate the fiftieth anniversary of full slave emancipation in Trinidad ran into opposition posed by a few white business leaders in Port of Spain, those on the other side had something more than hazy or abstract views of just what they wanted to celebrate and why. Looking toward the future, the last decade of the nineteenth century spawned a series of important workers' disturbances or riots—also marked by, among other things, incendiarism and torchlight parades—that were symbolic forerunners of the disturbances of the second and fourth decades of the twentieth century. These latter events, in turn, are usually considered by West Indian historians as direct precursors of political independence movements in the region.

The historically pivotal riots of the 1890s (and the related rash of sugarcane fires in Antigua and elsewhere) erupted for a number of reasons, perhaps the most important being the relentless economic depression that ravaged the area beginning in 1884, when sugar prices fell and would not rebound until well after the turn of the century.[14] In the summer of 1884 the dumping of an extraordinary volume of European, mainly German, beet sugar on London's open market drove down British Caribbean sugar prices, lowering local wages and food imports and creating economic misery in most of the islands where the production of raw sugar was the main economic activity. Workers' riots and disturbances were only some of the results of the depression. Malnutrition haunted the working classes on islands that had few economic alternatives to planter-imposed sugarcane production and where most food was imported. Some estate owners curtailed cane cultivation, which made local situations even worse. Eventually, in early 1897, a visiting commission from London traveled through the islands, seeking information, meeting with local officials as well as a few spokespersons from working-class groups, and eventually making some unprecedented recommendations. The commissioners suggested, for several of the islands, a breaking up of large land units into small-scale holdings. While these recommendations did not influence all the islands equally, and although planter oppression and influence continued, these ideas led to major changes in a few of the islands, allowing some of the working peoples of the region a measure of control over parts of their physical environments.

Other changes in the region were occurring as well. Within individual islands in the latter half of the nineteenth century, growing qualitative differences between cities and countrysides were heightening urban-rural distinctions. These differences were reinforced by the introduction of piped water throughout the region from the 1850s to the 1880s and by street lighting beginning in the 1870s. These amenities came first to the more important islands and first to urban places. Passenger railway systems appeared in Trinidad and Barbados early in the 1880s. Urban concentrations of amenities and activities, such as social clubs, lodges and other fraternal organizations, Saturday morning markets, and the presence of visitors from abroad, made towns and cities more exciting than rural villages. In early November 1889, for example, Professor Colby ascended an estimated 500 feet above the circus grounds of Port of Spain in a hot air balloon and then floated to earth under a parachute, an amazing event—witnessed by "a considerable crowd"—that simply could not have occurred in one of Trinidad's outlying villages.[15]

Similar urban-rural distinctions and experiences marked the smaller islands, too. The visit of Tony Lowandes's circus to St. John's, Antigua, in early 1903 featured a lion and an elephant, and the few rural Antiguans who could afford to come into town to see them would have had to do so by horse or donkey cart or on foot.[16] Far more imposing transportation limitations marked the outlying zones of the volcanic Windwards, where village communities on the windward or Atlantic sides of the islands were separated from the capital towns on the leeward sides by soaring volcanic peaks and knife-edge ridgelines, so that travel to the other side of one's home island was usually via sailing vessel. It was not unknown, despite these islands' tiny sizes, for old people of the windward villages never to have visited the local capital towns of Roseau, Castries, St. George's, or Kingstown during their entire lives.

Yet such isolation was exceptional, and increased connections were far more important within and among the islands. Younger people of the eastern Caribbean in the late 1800s were beginning to move from one island to another with more frequency than before, as opportunities attracted them or depression conditions pushed them away. These "migrations" were often temporary or seasonal, although some were permanent or nearly so, and all were being facilitated by improving modes of transportation. At the local level, wooden sailing schooners and sloops commanded by black West Indian seafarers carried an ever enlarging number of passengers and an increasing volume of produce from island to island for a variety of reasons. Also by this time, colonial officials routinely transported men from place to place, often as "deckers" on steamers, to fulfill various economic or colonial administrative objectives. Taken altogether, these interisland movements or migrations were causing the people of the region to view themselves and one another in new and progressively different ways. The mixing and intermingling were giving rise to incipient notions of "nationality" as well as to the stereotypes found in those same islands today. Talkative, self-assured Barbadians; mysterious, French Creole–speaking St. Lucians; and fun-loving yet violence-prone Trinidadians, not to mention men and women from elsewhere in the region (including those from non-British islands) were coming into contact with one another with increasing regularity.

Demographic enclaves of outsiders in all of the islands, especially in the larger locales, manifested this restless mobility and provided places to stay for itinerant friends and relatives who brought news from home. In a broader sense, the back-and-forth movement of people from place to place was forging a grassroots regionalism in the eastern Caribbean. It was not the same kind of formal "regional" framework devised by the men of the London Colonial Office who, referring to maps on the wall, devised multi-island political groupings for their own administrative convenience but that made little sense to the people living on the ground. Rather, it was a new and expanded habitation space, created by the movements and activities of the people of the region themselves and sustained by these same flows of people, goods, and information.

The unparalleled urban nucleus or node of this cultural region was Port of Spain, Trinidad, although every island had its capital port town from which travel took place and that was locally important in its own right. Besides being the logical focal point for group activity, an island's capital town or city also represented its official link to London and New York, and to neighboring places as well. All of these urban areas had originated on the sheltered leeward coasts in the days of sailing vessels, and now they housed the seats of local government and an island's main harbor and associated commercial settlement and port infrastructure, as well as the telegraph office connected to the undersea communications cable that had been laid in the 1870s. Thus capital cities and towns not only represented the communication and political nodes of particular islands, but they also served as windows on the outside world. Had an observer been unusually vigilant and astute from 1880 to 1905 in any of these British Caribbean port towns of the Lesser Antilles, he or she might have been able to discern—simply by noting the changing wharfside activities associated with official naval visits—the major geopolitical shifts that were occurring in the region at the time.

Early in the 1880s, such as in January 1884, when a fleet of six British naval vessels visited Antigua after stopping in St. Kitts, the ships of the British navy's North American and West Indian Squadron routinely anchored for a few days at each of the British island colonies of the Caribbean in the winter months. During these visits British naval officers and members of local elites renewed their acquaintanceships at planned receptions and competed in contests of cricket and rifle marksmanship. Knots of British sailors roamed the waterfronts, providing business for shopkeepers, fruit vendors, and prostitutes.[17] But as Britain's geopolitical interests and ambitions shifted and expanded elsewhere in the world, the visits by British vessels to their West Indian colonies became progressively less frequent, and by 1907 there was, mainly to the regret of the small number of white residents and officials in the islands, only a token and fleeting presence in West Indian waters of what had very recently been an enduring and formidable seaborne display of military might by the mother country.[18]

At the dawn of the twentieth century the British naval presence in the Caribbean was being supplanted by vessels from elsewhere, notably from the U.S. navy, which by 1899 "completely outclassed" Britain's North American Squadron in Caribbean waters.[19] At the macroscale and in retrospect, these changes are well known to historically oriented academics as the beginning of the "American Century" in the greater Caribbean region. Accordingly, it is not surprising that American ships began to visit the small British islands with greater frequency during this era. The three-ship task force composed of the *New York*, *Cincinnati*, and *Raleigh* from the American navy, as only one example, paid a series of courtesy calls in the eastern Caribbean in February 1895, stopping in, among other places, Castries, St. Lucia. There the American naval band accepted an invitation from the town council to perform at the bandstand at Columbus Square, providing late afternoon martial music that "delighted the large audience of all classes."[20] What neither the American sailors nor their West Indian audience could have known that afternoon was that three weeks later, the sailors from this American task force, including probably some of the band members, would be instrumental in helping to combat the enormous city fire in the center of Port of Spain, Trinidad.

The importance of the changing balance of military power in the Caribbean region at the turn of the century of course went far beyond the periodic observations of mildly interesting changes in the sights and sounds in the region's port cities and towns. So far as most local British officials and British planters in the islands were concerned, a withdrawal of the regular British troops, who had always been on hand or at least nearby to protect them from their own black laborers, was, to say the least, unsettling. When British troops were shifted from Barbados to St. Lucia in 1895 to provide greater security for the new coaling facilities in the latter island, white residents in Barbados wondered who would now protect them from a local populace "that has always been prone to rapine and riot."[21] One decade later the British governor of Trinidad articulated similar fears owing to the withdrawal of so many white soldiers and sailors from the region: "The excitable temperament of the coloured race has led to many disastrous disturbances, resulting in loss of life and dislocation of business."[22] It would have hardly been soothing to white Trinidadians or white Barbadians, or to the other white residents of the eastern Caribbean at the turn of the century, to know that their feelings of insecurity and fear and even terror would later be considered historically important sentiments that underlined and confirmed the momentous geopolitical shifts and dislocations occurring in the Caribbean region as the nineteenth century was ending.

2

Hazards, History, and Caribbean Fire

IN THE LATE AFTERNOON OF MARCH 4, 1895, A fire began on the top floor of one of the large wooden store buildings in downtown Port of Spain, the capital city of Trinidad and Tobago. Owing to seasonal dry conditions at the end of the Caribbean's low-sun period, abnormally high winds, and human inattention, what probably began as a storeroom blaze soon spread to other buildings and eventually expanded to become an urban holocaust. Yet the fire's destruction would have been far worse had it not been for the firefighting heroics of the British and American crews aboard four naval vessels that happened to be lying in the Port of Spain harbor on that afternoon. At dusk, when the fire already was raging, fifty British bluejackets from the HMS *Buzzard*, a destroyer that periodically patrolled the British Caribbean, came ashore to help. Shortly thereafter 250 American sailors and marines from a visiting U.S. naval squadron joined them. Together the British and American naval forces helped local firefighters to bring the worst of the conflagration under control. Then, during the night and early morning hours, armed marines from both countries patrolled the city's smoky streets and alleyways to help maintain security and prevent looting. Although the *Buzzard* stayed thereafter to help with security and cleanup, the Americans departed Trinidad, and a grateful British governor and officials, the next day.[1]

The joint Anglo-American effort in subduing the Port of Spain fire in

March 1895 was not the only multinational endeavor undertaken by rival colonial powers in the Caribbean region at the turn of the century. Germany and Great Britain, for example, carried out a joint blockade of Venezuela during 1902–3 in response to that country's debt nonpayment, an action also supported by Italy.[2] If events in a particular Caribbean locale could not be handled by local authorities, it was common at the time for them to request help from nearby vessels, regardless of their nationalities, to deal with the emergencies of natural or man-made hazards until proper local authority could be restored or reinforced. In August 1896 a British crew—using axes, crowbars, and hoses—led by Captain McKnight of the HMS *Andes* helped to prevent the spread of a major fire in the center of Port-au-Prince, Haiti.[3] In April 1907, only five years after the conclusion of the Boer War in South Africa, the British requested and received help from the Dutch man-of-war *Gelderland*, whose marines patrolled Castries, St. Lucia, during civil disturbances there.[4]

The Americans were especially capable and willing (and according to many Europeans, far too eager) to respond to Caribbean crises or calamities. At the turn of the century, the hazard-prone Caribbean region and the readiness of American financial, technical, and military might to come to the rescue represented the congruent pieces of a geopolitical whole that seemed made for one another. The huge San Ciriaco hurricane devastated Puerto Rico in August 1899, only a year after the island had come under U.S. control. The storm killed an estimated 3,000 individuals, caused US $20 million in property damage, and provided "an excellent opportunity for the United States to demonstrate its efficiency and supposed benevolence at a time of crisis."[5] Overall, the hurricane and its aftermath, while not directing the trajectory of Puerto Rico's future in a crude, deterministic sense, nonetheless provided an important environmental context for the earliest imprint and subsequent consolidation of American rule there.

The Trinidad firefighting adventure in 1895 was neither the best-known nor the costliest of the U.S. relief efforts directed to non-American territories of the Caribbean at the turn of the century. Far more extensive was the U.S. aid extended to relieve suffering from Mount Pelée's eruption in French Martinique in May 1902. The eruption monopolized newspaper headlines everywhere, and descriptions of the catastrophe itself—the estimated 30,000 deaths and the suffering that followed—captured imaginations worldwide. Within a week President Theodore Roosevelt and the U.S. Congress sent supply vessels to Martinique. The U.S. steamer tug *Potomac* called simultaneously at British St. Vincent with food and medicine to relieve the survivors of the eruption of Soufrière, the local volcano that flared up two days after Mount Pelée, killing 1,500 Vincentians.[6] In August of the next year, after a devastating hurricane destroyed crops and buildings on the northeast coast of Jamaica, the United States shipped food, directed local rebuilding efforts at the United Fruit Company complex around Port Antonio, and sent unemployed Jamaican men to Central America for banana work there.[7]

American disaster relief in Jamaica in 1907 sparked a major diplomatic controversy, an episode highlighting the growing U.S. domination (and associated European resentment) of the Caribbean region early in the twentieth century. In mid-January Kingston suffered a huge earthquake. Buildings collapsed, and 1,500 people died immediately from falling buildings or later from the citywide fire that followed.[8] Panic gripped local officials because the British naval squadron had been withdrawn the preceding year, leaving the city relatively defenseless against local unrest and, in this case, possible chaotic looting. The American naval detachment at Guantanamo Bay, Cuba, responded immediately to the telegraphed distress calls. Early on January 17, a three-ship American convoy arrived at Kingston.[9] But the next afternoon the British governor of Jamaica abruptly requested that the Americans withdraw because, as he put it, local authorities had control of the situation.[10] The governor's dismissal of the American task force, two days before the first British rescue vessel arrived, stunned local observers and stirred controversy on both sides of the Atlantic, provoking ambivalent commentary from Britain and decidedly anti-American reaction from both France and Germany.

The story of the Jamaica earthquake and its immediate aftermath—combined with the other Caribbean cataclysms of the era—highlights an environmental or geographical perspective on the beginning of the U.S. domination of the Caribbean.[11] The origins of the American Century in the region are invariably associated with the 1898 Spanish-American War as a prelude to the building of the Panama Canal.[12] This association has been reinforced by recent centennial observances and analyzed and portrayed strictly in terms of international geopolitical rivalries and power politics. Such an emphasis, furthermore, stresses—often to the exclusion of anything else—the economic and political ties between colonies and metropoles and among European nation-states and their representatives. Yet an awareness of the Caribbean's overall geophysical precariousness, not simply the seismic instability of southern Jamaica but also the region's hazard-prone geography, to include fire, provides an altogether different and, most important, locally based perspective that helps us understand, among many other things, how the Americans came to dominate the Caribbean early in the twentieth century.

The hurricanes, earthquakes, and fires that ravaged the Caribbean at the turn of the century were certainly nothing new. The entire region, including the Greater Antilles, is seismically active. The string of tiny, mountainous islands composing the volcanic arc of the Lesser Antilles marks the dynamic intersection of two of the earth's crustal plates. Damaging earthquakes and occasional full-blown volcanic eruptions have therefore occurred throughout the region's history, and they will occur there in the future. More frequently, hurricanes and tropical atmospheric depressions, because they are of a seasonal nature, relentlessly punctuate every late summer season in the Caribbean region. From June to November these huge storms proceed west from the equatorial waters off West Africa, and although hurricanes' seasonal occurrences are predictable, the storms' individual trajectories—beyond the generalization that they usually move from southeast to northwest—are not.

The immediate impacts of the Caribbean's recurring hazards, while drawing the fleeting attention of newspaper readers and television viewers elsewhere, obviously are felt most directly at the local level. Furthermore, the region's distinctive insular geography plays a crucial role in illustrating how thoroughly devastating geophysical hazards are now and have been in the past. Tiny islands are transformed more readily than are continental landmasses, a point more profound than it is obvious throughout Caribbean history, and the coupling of small island size with geophysical cataclysms has often led to rapid and dramatic change. An island's entire sugarcane crop, and sometimes its entire forest, has been destroyed by a single hurricane. When a hurricane or a volcanic eruption or a major fire destroys the only real town or city of an island, as in the case of Kingston in 1907, St. Pierre, Martinique, in 1902, or Port of Spain, Trinidad, in 1895 (not to mention the igneous burial of Plymouth, Montserrat, late in the 1990s), the only real recourse for socioeconomic recovery is to seek external relief.

Important long-term decisions and turning points in Caribbean history—beyond those determined by the actions of people coping with the immediacy of a single catastrophe—also have been influenced by the region's hazardous environment. Fundamental economic change in Cuba resulted from the influences of three particularly devastating hurricanes in the middle of the nineteenth century.[13] The widespread publicity of the Mount Pelée eruption combined with reports of simultaneous eruptions in Nicaragua was an important reason the United States decided in 1903 on Panama, not Nicaragua, as the locus of its trans-isthmian canal effort.[14] The rapid transformation of St. Vincent from a sugarcane island in the 1880s and 1890s to an island of smallholders in the early twentieth century probably was brought about by an 1898 hurricane followed by a volcanic eruption in 1902, twin disasters that drove away most of the island's planter class.[15]

Could the Caribbean's natural catastrophes even have diminished planter resistance to slave emancipation itself? According to Lowell Ragatz, the local destruction caused by a series of massive and violent hurricanes in the British Lesser Antilles in the late 1820s and early 1830s "found the planters . . . in such a weakened state that . . . many . . . were obliged to abandon their estates because of being unable to repair the damage done."[16]

The aperiodic, unpredictable nature of the Caribbean's recurring hazards, combined with its fragmented insularity, provides an important and dynamic geographical backdrop for understanding a region whose human history consists of a series of pivotal discontinuities. Persistence, tradition, and collective memory count for much in the Caribbean, and one can make a strong case that past and present are more closely bound up with each other there than in some other parts of the world. But the Caribbean's past is not a gradually changing, linear, or seamless stretch of time. Rather, it is characterized by a sequence of sharp disjunctures that have then ushered in unprecedented change; first human habitation, European contact or "discovery," slave emancipation, and political independence are perhaps the most important of these watershed events, temporal fault lines that are second nature to every student of Caribbean history. Much of the reason for the punctuated pattern of its past is that the Caribbean has been colonized for nearly all of its recorded history. Decisions made and actions taken elsewhere subsequently have been imposed on and have transformed abruptly various parts of the region, and the region in its entirety, more than once.

Yet the Caribbean's physical characteristics have helped to influence, shape, and facilitate those decisions as well. Prevailing easterly winds brought sailing vessels from Europe whose crews afterward headed for higher latitudes where they found they could ride the westerlies home. Later, small Caribbean islands provided little physical refuge for aboriginal peoples against the onslaught of European invasion, and small island size allowed easy seaborne accessibility for subsequent European sailing vessels bringing cash crops and African slaves. Further, tiny insular land areas, owing in part to their smallness, could be transformed from multispecies natural forests to monocrop economic units in relatively short order. Thereafter, the region's hazards could hasten planter decisions to quit particular islands in the face of declining crop revenue. A string of neglected, worn-out, economically depressed island colonies, beleaguered by a series of recurring geographical catastrophes, could provide—at the end of the nineteenth century—an ideal opportunity for a burgeoning, nearby political power to demonstrate its technical proficiency and economic largesse by coming to the rescue.

A history of the Caribbean region that takes into account the importance of the area's major hazards carries with it the virtue of emphasizing grassroots events and considering the way those living on the islands have coped with and thought about recurring environmental problems. The 1899 San Ciriaco hurricane, for example, likely was the single most memorable event in the lives of an entire generation of ordinary Puerto Ricans, just as for young Montserratians of the late 1990s, "the generation of children who saw and heard the volcano . . . will be the ones to tell their volcano stories to their children."[17] Similarly, historians can never know whether or not the residents of St. Vincent knew (or cared about) the names of the various government officials of the British Windwards from 1898 to 1902, but they can be certain that every resident was acutely aware of the 1898 hurricane as well as the volcanic eruption there four years later.

Whenever Caribbean hazards occur, they always inspire the retelling of memories of similar events. A few very old people on St. Vincent in 1898, for example, were able to recall—after the huge September hurricane of that year—past storms, even back to the Great Hurricane of 1831.[18] These recollections, now and in the past, and the similar remembrances that invariably accompany the region's recurring catastrophes are far more than colorful anecdotes about the old days. They provide local depth and perspective about the precarious environmental underpinning of a region where earth tremors, sudden storms, and the turbulent socioeconomic changes often associated with these experiences—including large fires—seem more the rule than the exception. Living memories, and in some cases written records, of these recurring events and their innumerable consequences thereby reinforce the continuity between the present and the past and influence the way Caribbean peoples regard the future.

Caribbean Fire in History

Large fires represent momentous environmental experiences on small Caribbean islands, and like earthquakes and hurricanes, they are also recurring events. Further, sizable fires stimulate their witnesses and victims to reflect upon the comparative effects of earlier conflagrations in local history. The March 1895 Port of Spain fire provided an excellent case in point. Before the smoke even cleared, officials were comparing it to the city's great Frederick Street fire of 1859 (which, most agreed, probably was less damaging). Those taking an even longer view insisted that the 1895 fire was in many ways comparable to the city's immense conflagration of March 24, 1808, a fiery disaster that had begun very close to the recent one and "which in a few hours swept away nearly the whole Town."[19] Recollections of the 1808 fire, further, provided vivid and useful examples for Trinidadian authorities in 1895 who emphasized the close similarities between fires and natural catastrophes in terms of human suffering. When the terrible September 1898 hurricane devastated Barbados and St. Vincent and then swerved north to graze St. Lucia, the *Port of Spain Gazette* recalled for its readers how generous residents of those nearby islands had been to Trinidadians who had suffered from the fire in 1808, a reminder intended to publicize the obvious need for Trinidad's residents to reciprocate one century later.[20]

Given the flammability of the wooden buildings in the towns and cities of the eastern Caribbean of the 1880s and 1890s, even hearsay about fires in neighboring places provoked dread. News of the large fire in late June 1890 that burned down Fort-de-France in Martinique stirred deep emotion throughout the British islands nearby. Part of the feeling was genuine sorrow for the 5,000 homeless Martinicans, but probably more was a realization that similar fires could happen at home. Reports of the Martinique fire hit particularly hard in St. Lucia, where many residents had friends and relatives in the nearby French island. More immediately, St. Lucia's capital, Castries, was made of wood (and kerosene was used for lighting and cooking) and was portrayed in the local newspaper as frightfully combustible. According to the newspaper, much of the "depth of emotion" that the news of the Martinique fire had created in Castries arose from the fact that "we are exposed at any moment to the same or a worse fate, from which extraordinary good fortune has alone, until now, saved us. . . . There is no organisation or appliance to oppose to a conflagration, and we could not get up a head of water to reach a first story window to save our lives."[21]

The threat and hazard of large Caribbean fires, whose destructive capacities were universally acknowledged as similar to catastrophes wrought by nature, nonetheless involved an entirely different element. With rare exceptions, fires in the region were and are attributed to human agency. The environmental or inanimate side of the fire equation in the form of fuel was obviously crucial, and an environmental or seasonal dimension, most notably in terms of dryness that promoted combustibility, was important. But ignition was nearly always human induced. So the written reports and accounts of Caribbean fires at the turn of the century had a hectoring, accusatory, and indignant tone, wrathful rhetoric that was completely absent in reports of earthquakes or hurricanes, which were seen more in terms of hopeless resignation. The latter were acts of God. Damaging fires, on the other hand, whether accidental or purposeful, were ultimately attributed to the acts of humans, who should therefore be cautioned, educated, or blamed.

Fire in the region also took on innumerable forms and functions, not all of which were dangerous or unhelpful. Cooking fires were obviously ubiquitous. Kerosene lamps lighted homes and, in some cities and towns, the streets. Smoldering trash heaps marked the shanty towns of the larger urban areas. Rural dwellers routinely burned farm refuse. Smoky fires drove away mosquitoes and other insects. Mountainside charcoal burners played useful, if usually illegal, roles in insular economies, as did the illicit producers of rum, whose activities often were detected by wisps of smoke. Boiler fires at sugar estates converted sugar juice into molasses and raw sugar, and by the turn of the century, preharvest burning had become integrated into sugar-cane cultivation routines in some of the islands. Fire, in other words, played integral and varying roles in daily life, but since it could transform itself, in short order, from being useful to becoming dangerous, fire—and those using it—needed control.

The control of fire and the associated control of people in the fin-de-siècle British Caribbean involved rules, domination, and legal prescriptions that had their roots in the region's plantation past and which, at least indirectly, reinforced memories of slavery. The tiny white populations still initiated, legislated, and enforced local laws in the islands in the late 1800s, and they doubtless considered local legal rules and regulations to be extensions of a broader British civilization that promulgated a system of impersonally prescribed codes of conduct intended to benefit and protect one and all. The black majorities in each place, had their opinions ever really been sought, almost certainly would have condemned many of the same laws as thinly veiled oppression and an extension of slavery that was, after all, the legally sanctioned institution that had brought their ancestors there in the first place. During slavery, planters had owned everything—people, land, buildings, and implements—a legality that usually could be enforced for tangible, fixed objects yet was nearly impossible for fire. Perhaps this is one of the reasons why Caribbean resistance and rebellion, imagined and real during slavery and thereafter, were nearly always associated with images of burning fields or buildings.

For whatever reasons, British Caribbean fire—especially fire that was out of control or, worse, controlled by the wrong people—at the turn of the century invariably frightened local officials and others with vested interests. Nor was this fear simply because of fire's capacity for destruction. Fire could simultaneously assemble a crowd, reawaken enduring animosities, resurrect the past, and remind observers of the region's stark socioeconomic inequities. Groups of black men brandishing torches, either in protest or for ritual display, was activity that needed tight control or, better, legal prohibition so that matters would not get out of hand. Even if burning started accidentally, fire could evoke fundamental fears rooted in the past. Shortly after the eruption of Mount Pelée in Martinique, a story circulated about a young white boy there who had been saved from the catastrophe by an older black man. The St. Lucia newspaper extended this anecdote to tell of an earlier slave insurrection in southern Martinique in which escaped slaves pillaged and burned, an article that doubtless struck a responsive chord among anxious white readers in 1902 who harbored fears that history easily could repeat itself. Stopping at one plantation house, according to the story, the Martinican slaves ruthlessly killed the estate owner, whose wife and daughter had earlier been secretly led away by a faithful black servant to hide, terrified, in a nearby cane field. "By the light of their flaming home they could see their pursuers, and hear their cries of baffled rage."[22]

Reports of lesser fires at the time also illuminated prevailing racist attitudes. Newspaper descriptions of local blazes often included gratuitous observations of "savage behavior" by those who cavorted mindlessly around burning buildings or sugarcane fields, thereby exhibiting loathsome acts of "African barbarism." The occasional reports of Haitian fires in British West Indian newspapers, furthermore, seemed always to delight in providing sensational word pictures highlighting the supposed anarchy prevailing there. An article in the *Port of Spain Gazette,* as one of a number of similar examples, regaled its readership with "The Condition of Haiti" in July 1888. Early that month, purposefully set, incendiary fires had burned hundreds of dwellings and businesses in Port-au-Prince. French sailors, whose vessel happened to be in port, provided general firefighting help and then protected both the British and French consulates from looting and the "hopeless chaos" reigning in the city: "Other fires are expected to take place hourly. . . . Two men who are supposed to be crazy have since been arrested and publicly shot for being the persons who had fired the town. . . . The greatest confusion prevails all over the city and business is entirely suspended."[23]

British planters' and officials' subliminal fears of turn-of-the-century social upheavals that might be ignited or spread by Caribbean fire were of course predated by several millennia by the first human use of fire in the region. Aboriginal burning throughout the Caribbean had preceded plantations and sugarcane and was far earlier than European entry into the region. Indeed, Christopher Columbus, it may be recalled, anticipated his immediate landfall somewhere (he thought) near Asia with the possible sighting of a flame. Before moonrise on the night of October 11, 1492, Columbus, from his position on the sterncastle of the *Santa María,* thought he saw a light in the far distance. He was uncertain about what he saw, yet it momentarily seemed to him "like a little wax candle rising and falling."[24] Subsequent historical interpretations of Columbus's possible sighting of firelight on that night have varied to include the idea that he might have seen torch-bearing aboriginal fishermen trolling the reefs and shoals off the Bahamas.

Since aboriginal West Indians, like all people everywhere, marked their activities and presence by fire, archaeological evidence of burning plays a crucial role in studies attempting to date the first human entry into the region. The earliest presumed human occupation of the Caribbean, roughly 6,000–7,000 years ago, is of people who rafted east from Central America to inhabit Cuba, Hispaniola, and possibly Puerto Rico. Later groups—the better-known Arawaks, followed by the Caribs—then island-hopped from northern South America, populating, or perhaps repopulating, the Lesser Antilles from as early as 2000 B.C. in Trinidad to as late as 400 B.C. in the Leewards.[25] Recent interpretations of charcoal deposits in Haiti may validate the earliest human presence in the Greater Antilles, although whether the 7,000-year-old fires there were of human or natural origins are subject to intense debate.[26] Recent analyses of charcoal evidence from much later aboriginal sites in St. Eustatius and Montserrat suggest the possibility of a wider variety of cultivated plants and native flora than earlier suspected.[27] The coupling of fire evidence with earliest human activity in the region, it seems important to point out, is complicated by the Caribbean's hazardous physical geography. Hurricanes are well-known agents of vegetational change that can transform untouched forests anywhere into piles of combustible timber that are then subject to drying and ignition by natural means; "undoubtedly similar events occurred in the prehistoric Antilles."[28]

At the height of their presence in the Caribbean—immediately prior to the arrival of Europeans—aboriginal West Indians used fire for household activities, hunting, cultivation, and ritual. Ignition was accomplished by spinning wooden drills between two sticks within a combustible, tightly wrapped bundle, thereby producing fire from friction. Aboriginal ceramic work required baking ovens and drying kilns. Both Arawaks and Caribs cremated bodies and preserved the bones of dead kin. Cooking techniques depended on fires of varying sizes, to include the use of "the *barbacoa*, a grate or trivet on which the meat was dried over a fire." Geographer Carl Sauer reported that the earliest Spaniards on Hispaniola told of aboriginal fire hunting on both land and sea. "Fire drives" of rodents and hares toward waiting hunters involved the selective, seasonal burning of savanna grasses. Also, "canoes went out at night carrying torches to attract the fish, which were then harpooned or shot by bow and arrow."[29] Even the means of inter-island travel depended, strictly speaking, on fire. Aboriginal West Indians sailed from one island to another on large, painted canoes (*canoa*), some of which could hold up to 150 people. The large vessels were "hollowed out of logs by alternately charring and chopping them with . . . stone axes."[30]

Throughout the region, aboriginal cultivators used the slash-and-burn system of forest farming, in which vegetation at preselected sites was chopped down (or girdled), allowed to dry, and then fired at the end of the dry season. This burning technique eliminated most of the felled logs and branches and released mineral nutrients into the soils through the resulting ash.[31] Then, varying combinations of crops—notably corn and especially cassava—were cultivated in the newly cleared patches. Yet the extent to which Arawak and Carib slash-and-burn agriculture modified overall forest covers on the islands of the region may have been slight. The Arawaks grew their cassava in *conuco* tillage, the term associated with mounded piles of earth that represented an intensive, somewhat permanent character to their cultivation and which did not require the continual movement, arduous clearing with stone axes, and burning so often associated with "shifting agriculture" elsewhere. Insular forest covers may also have been protected indirectly by aboriginal preferences for the sea for subsistence; recent reviews of pre-European archaeological sites in several different islands suggest "that with time, marine resources became increasingly important to West Indian horticultural peoples."[32]

Records of the first sightings and initial impressions of the islands by Europeans provide subjective confirmations of relatively undisturbed forest covers. Yet one must keep in mind that the comparatively cutover character of more familiar Mediterranean islands at the time may have influenced the European sailors' observations. "Espanola is a marvel" enthused Columbus, who described a place "filled with trees of a thousand kinds and tall, and they seem to touch the sky." He also heard that the many varieties of trees ("as green and lovely as they are in Spain in May") never lost their foliage.[33] On the second voyage, Columbus's crew acquired firsthand experience of the impenetrable character of some of the region's vegetation when ten of them became lost in the forest in Guadeloupe and a search party sought them for a week before they were returned to camp by local peoples. Thereafter, as he continued sailing northwesterly toward Hispaniola, Columbus witnessed variety in how aboriginal peoples influenced local forest covers. Passing Santa Cruz (St. Croix), said to be a Carib island by his Arawak informants, the Spanish crew saw that the island "was intensively cultivated and gave the appearance of one large garden as they cruised along its shores."[34]

Whatever the aboriginal flora of the region, European entry soon changed it irrevocably. Peter Hulme describes the social impact of 1492 as an irreversible event: "The European invasion . . . shattered the socio-political evolution of Caribbean societies, destroying for ever (or better, taking over and thereby destroying) the established chiefdoms."[35] This sharp sociohistorical discontinuity had its ecological accompaniment in the European-induced destruction and transformation of the natural vegetation of the Caribbean realm, a dramatic environmental change accomplished by clearing and burning. The first massive destruction occurred in the Greater Antilles at the hands of the Spanish, who in the first decades of exploitation obliterated the aboriginal population in their quest for gold and transformed much of the environment. In Carl Sauer's oft-quoted summation, "By 1519 the Spanish Main was a sorry shell."[36]

Ecological transformations similar to those in the Greater Antilles at the juncture of the fifteenth and sixteenth centuries occurred in the Lesser Antilles one century later. In the first half of the 1600s, Dutch entrepreneurs introduced sugarcane via northeastern Brazil to the tiny French and English settler colonies of the eastern Caribbean.[37] Within a few decades, what had been small-scale European subsistence-, indigo-, and tobacco-farming footholds at the edge of the Caribbean Sea became market-oriented sugarcane islands populated largely by imported slaves from West Africa. The latter replaced aboriginal peoples whose numbers were greatly reduced or who were killed off completely by warfare and imported disease. Overall insular floral assemblages were similarly transformed; cash crops were planted after the clearing and incineration of the natural vegetation, a massive task given the technology of the era. Sun-loving weeds then invaded cleared areas where forests once stood and encroached on the borders of the wooded areas left standing on the steeper slopes. Without natural forest covers to ameliorate periodic dryness or to intercept and absorb intensive rainfall, both droughts and floods became more common in the islands. Massive soil erosion followed, especially on the steepest slopes. In short, the seventeenth- and eighteenth-century clearing of the Lesser Antilles to make way for sugarcane represented a pivotal discontinuity in the economic and ecological history of the Caribbean, a process that would have been impossible without fire.

The archival reports of the earliest European encounters with the Lesser Antilles tell of the difficulties in clearing the tough tropical vegetation that extended to the water's edge. The thick trunks, dense foliage, and myriad vines represented formidable hindrances to settlement and even to human locomotion, so that the settlers had little recourse except to resort "to the easy but very dangerous course of burning the jungle."[38] Yet despite the dangers, burning soon preceded clearing itself. So earliest European cultivators in the islands had to intersperse their planted crops "Arawak-style" in the openings among charred stumps and rotting logs, obstacles they never had encountered before. In the words of Sir Henry Colt, who visited Barbados in 1631, the local landscape "lay like the ruins of some village lately burned, here a great timber tree half burned, in another place a rafter singed all black. There stands a stub of a tree about two yards high, all the earth covered black with cinders, nothing was cleared."[39] On Martinique, European-set fires modified the vegetation before permanent European settlement there. Although most Martinican planters tried to use fire responsibly in land clearing, it sometimes got away, so by the 1660s, "one could pass easily to all parts of the island on horseback, suggesting perhaps an opening up of the floor under forest canopies by fire."[40]

As Europeans proceeded to the more northerly and therefore drier islands of the eastern Caribbean, forest removal must have been progressively less difficult—though it was described as arduous and time consuming everywhere. In the north the greater incidence of drought and xerophytic vegetation would have made these places somewhat easier to burn than those farther south.[41] Geographer David Harris notes that among the various methods of clearing the original vegetation in the Leewards—including Anguilla, Antigua, and Barbuda—by first European settlers, "fire was probably the most potent agent of ecological change." He further points out that African slaves were first brought to Antigua by the English so that they could be set to work immediately, clearing and burning the tangled scrub in lower elevations of the island and the forested zones at intermediate heights so as to render the place healthier and more fit for European habitation. By 1676 African slaves had incinerated sufficient vegetation on Antigua so that the sun could penetrate the dank interior recesses and allow the island to "exhale those vapors" that were thought to lead to disease and ill health.[42]

The earliest European, or European-directed, clearing and burning of Caribbean forests was thus not simply for the utilitarian removal of wild vegetation to make way for towns, roads, and cultivated plants. In concert with the clearing for plantations and crops, it was common, given the state of scientific knowledge of the time, for native vegetation to be removed so as to eliminate the unhealthy "fevers," "vapors," and "airs" that hung in early morning mists or lurked in the soil waiting to escape. Since Europeans were traveling more extensively to new and unknown places, European medicine was taking closer accounts of obviously varying disease environments in different climatic realms. As Europeans established themselves in these new places, the strange surroundings themselves often were held accountable for sickness. About 1642 the Dutch established a foothold on Tobago. "At first they found the climate sickly and unwholesome, yet by degrees, as they cleared it, the air agreed with them better, and they began to extend their settlements."[43]

Given this level of environmental misunderstanding, an obvious remedy for alleviating maladies was to modify local surroundings as the Dutch had done on Tobago. By extension, one can envision and comprehend better the attitudes and actions of colonial ships' crews of European men in alien and frustrating new places, far from any real supervision, and perhaps experiencing illnesses of varying kinds. These men came armed with vague yet authoritative scientific hearsay that blamed alien and uncomfortable physical environments for ill health, and they also possessed the means to create fire. Thus burning—for medical reasons and as ill-tempered outbursts against physical and social circumstances they simply could not understand—may have come naturally and spontaneously to them. The most astounding example of ecological destruction in the early colonial Caribbean to "improve" health conditions may be a fire described in a report from about 1650. A shipload of French sailors contemplated landing on St. Croix yet noted that "the island was covered with old trees, that prevented the circulation of air. To remedy this inconvenience and make the island more healthy, the French set fire to the woods, and from their vessels watched its progress: it burnt for months." When the fires subsided, they landed again, after their burning presumably had rendered the island more fit for human habitation.[44]

Although begun by Europeans, the backbreaking chopping and burning of the native vegetation of the Lesser Antilles soon became work carried out, almost certainly exclusively, by African slaves. It is also little wonder, given the exertion that forest clearance must have required, that the annual death rates during the first decades in which slaves were brought to the region were so tragically high. Geographer David Watts suggests that the deforestation of central Barbados from about 1645 to 1665 involved extraordinary human effort with unimproved iron tools devoted to a labor objective that saw "stumps levered out of the ground and burnt . . . to make way for sugar estates." He concludes that the results of this forced labor were profound, "even to the point of death in many instances."[45] Watts's grim speculation can perhaps never be confirmed, except by the general correlation that the always extraordinarily high slave mortality rates in the Caribbean region were even higher in the first (deforestation) decades.[46] Slave "seasoning," the year or so during which new slaves often were acclimated to the rigorous work routine of British Caribbean islands, seems not to have come into practice until the eighteenth century, after most of the native forest cover was cleared away.[47] In any case, it seems important to point out that the removal and firing of much of the natural vegetation of the Caribbean islands was an ecological transformation likely accomplished at the cost of thousands of African lives.

Fire and the African entry into the Western Hemisphere, at least in the early decades of Caribbean slavery, were probably bound up with each other even more closely than previously thought. In the latter decades of the 1600s and into the next century, as sugar profits ran high and the demand for new cane land soared, chopping and ring-barking of trees could be carried on in the islands year-round. The burning season, on the other hand, almost certainly occurred in the relatively dry first half of the calendar year. The smoke from these fires, and the smoke lingering thereafter from charred logs and smoldering stumps, must have hung over the islands seasonally, creating hazy atmospheric conditions that island residents came to expect. At the same time, the introduction into the Caribbean of at least some of the new African slaves (to replace those who had died), whose presence was vital in order to facilitate and power this momentous environmental change, also would have come during the smoky season.[48] It is therefore a near-certainty that among the first sensations that some African slaves had of the Caribbean region, as they stumbled from the holds of slave vessels, were sights of smoky skies and smells from local clearing fires, initial sensual impressions that were in many ways accurate precursors of what the inferno of Caribbean slavery would be like.

Fire literally imprinted many, if not most, African slaves before they arrived in the Caribbean because of the practice among the slavers from several European countries of branding with hot irons on shoulders, chests, or arms.[49] In the islands themselves, apparently little information was recorded of the domestic uses of fire within slave households, although cooking fires had to be ubiquitous. At the end of the slave era in some places, slave meals came from central plantation kitchens. Excavations on Barbados suggest that slaves cooked their own food by roasting or boiling it over open fires outside individual dwelling places, using stones to support iron pots placed over the cooking fires.[50] In Jamaica slaves lit their dwellings with torches, thereby increasing the danger of accidentally burning down their tiny huts and cabins. Construction material for slave dwellings varied throughout the islands, although many were of thatch placed over wooden frames. It is therefore unsurprising that the combustible slave houses routinely went up in flames, as they were "constructed on a skeleton of wood . . . easily ignited and quick to burn."[51]

The more substantial buildings of the towns and cities in the slave era also were susceptible to destruction by fire. A "considerable part" of St. John's, Antigua, lay in smoldering ruins after the fire of August 1769, and then accidental fire revisited the town in April 1782, attacking in particular its wooden house shingles "dried by the scorching rays of a tropical sun."[52] Earlier in the eighteenth century, raids back and forth between rival European naval powers vying for control of individual sugar islands were often devastating because of the destruction from burning. The precariousness of planters' lives and livelihoods on Nevis early in the 1700s was underlined by occasional raids whose resultant destruction and damage could be "felt for a generation. . . . The flimsy timber buildings would burn like tinder, and so—which was worse—would the canes."[53]

During the eastern Caribbean slave era, fire was an omnipresent element and a symbolic foundation for the production routine on every sugarcane plantation. At harvesttime the cut canes were crushed for their juice, an operation accomplished in the earliest decades at animal-powered roller mills and later—in the late 1600s in Barbados and in the early 1700s elsewhere—at wind-powered grinding factories.[54] The juice then was transformed into molasses and raw sugar at the adjacent boiling houses that, with the mills, were the logistical focal points for every plantation. Details of the boiling operation varied from place to place, but the general pattern was the same. Sugar juice flowed via stone gutters into a series of fire-heated iron containers or pans, each of which represented a step in the sugar extraction routine. At one point the juice was "struck" when certain chemicals were added. At another, impurities were skimmed from the smoking liquid. Finally, the molten juice was cooled, usually in earthen pots, and destined to become molasses or raw sugar. The necessarily small plantation sizes, roughly 200 to 300 acres each, facilitated the transportation of the cut canes to and their subsequent crushing at each mill within the forty-eight-hour period before the sucrose content began to fall. So the combined grinding mill–boiling house complexes of different estates were reasonably close together and nearly always within sight of one another, and at night during the January-to-June harvest season, anyone looking across the newly cut sugarcane fields would have observed the many scattered lights of fires from several neighboring boiling houses, not just his own.

At harvesttime, fires burned around the clock in the sugar boiling houses, adding to the sweltering heat during the day and providing flickering light after dark. The rollers of the nearby mill often turned all night—the factory's wooden scaffolding creaking and groaning as the canvas sails turned—in order to produce the sugar juice. As the liquid flowed into the boiling house, slave technicians and boilermen used heavy, long-handled wooden ladles and skimmers to direct its passage into the line of smoking sugar juice containers. Stokers removed ashes, and they tended and fueled the continuous fire "as if it were alive." The ambience of the boiling house was close to suffocating, and it was always a place of perpetual noise and motion. The so-called satanic mills of the Industrial Revolution in Europe were not so much duplicated as they were preceded by the furnacelike sugar boiling houses that smoked and sweltered at the center of every Caribbean sugar plantation.[55]

In the earliest plantation years, fuel requirements for boiling houses went hand in hand with insular deforestation, and the cut timber was often hauled directly to the estate nucleus, where it could dry before it was burned. The dried forest timber, combined with local bush and scrub branches, fueled ideally hot fires. But finite local wood supplies dwindled over time, and further forest removal on steep slopes became impractical and even dangerous. Occasional "free freight" coal shipments from Britain, brought by ship captains who required ballast for otherwise light loads outbound for the islands, sometimes augmented local fuel supplies.[56] But the eventual solution to a growing shortage of fuel for boiling was to substitute the crushed cane itself, the dried fibrous substance that came to be known as "bagasse," for firewood. This innovation, in turn, necessitated better cane crushing techniques that would produce drier, and therefore more combustible, bagasse. The burning of crushed sugarcane bagasse for boiler fuel first became common in the Lesser Antilles, where timber supplies were obviously smaller than on the larger islands, by the late 1600s. Thereafter, bagasse houses or "trash houses," where dried sugarcane stalks were stored as fuel, became standard landscape elements of the sugarcane plantation complexes throughout the region.[57]

So by the early 1700s in nearly all of the tiny British colonies of the eastern Caribbean, agro-industrial sugarcane landscapes had replaced what had been only slightly modified natural vegetation just a few decades earlier. In the lower reaches of every island, below the wooded highlands, a ten-foot-high forest of sugarcanes—altogether representing millions of tons of biomass—grew for twelve- to eighteen-month periods before it was cut, crushed, and burned, and then the cycle was begun again. Prior to the coming of the Europeans, rainfall, soil nutrients, and solar energy had combined, more or less undisturbed by human intervention, to sustain complex, multispecies forest ecosystems. But sugarcane plantation owners now owned these islands, and they essentially harnessed these same ecological elements into an entirely new, cyclical biological system of rapid plant growth punctuated by abrupt seasonal plant removal, a sequence repeated year after year, and an agricultural system monitored and directed by men whose ever-present clearing agent and catalyst was fire.

But fire as agent and catalyst could quickly become fire as menace, because slave resistance and burning fields—in the minds of probably both masters and slaves—always went together. After slaves on Nevis were denied the right to grow cotton for sale in 1777, they were suspected of resorting to arson as a form of protest. In the following year the island's legislature "passed a law inflicting death upon any person found guilty of wilfully setting fire to any building or cane field."[58] At a far broader scale, fire was a feared symbol of the possible massive overturning of an entire plantation system, a prospect that by 1800 must have been dreaded by every Caribbean planter because of the horror of the Haitian revolution. If slaves grumbled or were more recalcitrant than usual in the early 1800s, or if general discontent was even rumored, Haiti must have come to mind. The stories of murdered French planter families, general pillage, and rampant looting were bad enough, but it may have been the stories of burning that were the worst of all. The conflagration at the start of the Haitian revolution, in a description probably similar to what passed orally from one British Caribbean planter family to another at the time, was captured by C. L. R. James in one of the most vivid passages in all of Caribbean history: "In a few days one-half of the famous North Plain was a flaming ruin. From Le Cap the whole horizon was a wall of fire. From this wall continually rose thick black columns of smoke, through which came tongues of flame leaping to the very sky. For nearly three weeks the people of Le Cap could barely distinguish day from night, while a rain of burning cane straw, driven before the wind like flakes of snow, flew over the city and the shipping in the harbour, threatening both with destruction."[59]

All of the slave uprisings in the eastern Caribbean that preceded emancipation in the 1830s involved the use or planned use of fire. In the aborted plot to take over Antigua in October 1736, the slave conspirators in all likelihood planned to kill the leading planters of the island by exploding gunpowder at a prearranged planter gathering and then falling on the assembled whites "with fire and sword."[60] At the end of the eighteenth century, during the French-inspired Fédon rebellion in Grenada, panicked planters fled the chaos and conflict in the countryside by heading to the capital of St. George's, only to have their plantation houses looted and burned. What has come to be known as Bussa's Rebellion in Barbados in April 1816 lasted for only a few days, although martial law endured for months thereafter. The organized slave revolt, centered in St. Philip parish, was planned for Easter night, when a fire of bagasse and cornstalks on River Estate would signal others to rise up. Other fires were to follow, with the uprising's overall objective "to set the country alight." The first fires of rebellion were indeed lit as planned, and within hours fires spread contagiously to the adjacent sugarcane plantations of the southeastern part of the island. By early afternoon on Easter Sunday, April 14, 1816, on Barbados, "a third of the island was aflame"[61]

Yet countless daily acts of resistance to slavery carried out on a routine basis—balking, malingering, feigning illness, and so forth—were far more common than open revolt. By the early 1800s the tiny sugarcane plantation colonies of the eastern Caribbean, an arc of cultural landscapes in many ways created and sustained by burning, provided the settings for an overt and recurring form of ritualized resistance involving the direct use of fire, an active celebration of liberation and aggression that endured for decades and lasted until well after emancipation. The countryside's many sugar boiling houses and, by the early nineteenth century, innumerable bagasse storage shelters, along with the cane fields themselves, were important yet flammable elements of the local sugar economies. Unplanned and accidental plantation fires were sufficiently common that planters routinely dispatched slave firefighting gangs to nearby estates to help extinguish the blazes. At night these mobile gangs of slave firemen illuminated the roads and paths by carrying torches, their treks thereby taking on the form of a parade. Although these activities were, strictly speaking, under planter supervision, the otherwise forbidden torchlight processions must have had an unshackled, liberating, somewhat illicit, and even raucous character about them, very different from the supervised monotony of daylight toil. These firefighting ventures apparently were sufficiently exciting so as to become discussed, even anticipated, among the plantation slaves. This story, in any case, was the one black freedmen and their families throughout the region told to one another after emancipation in accounting for the origins of the pre-Lenten *cannes brulées* torch parades. This playing out of the memories of slavery, defiantly symbolized by recurring and ritualized *cannes brulées* processions in the decades after emancipation, frightened and antagonized local whites and colonial officials in Trinidad and nearby islands into the late nineteenth century and beyond.

Changing Regional Identities in the Late 1800s

The growing distinctions between urban and rural in the British islands of the eastern Caribbean during the late nineteenth century were obvious to all, so they influenced everyone in the region, city and country peoples alike. Because each island was so very small, what happened in "town" was usually well known to everyone within a brief time, although the speed and nature of the diffusion of different kinds of information obviously varied from one island to the next. Rural agricultural estates, to complicate matters further, often were the loci of the most modern and up-to-date social and economic activities, as they usually housed the latest agricultural machinery and equipment. In any case, the rural village areas of the Lesser Antilles, although impoverished and behind the times in outward appearance, did not necessarily house an inert, unchanging peasantry distant from centers of innovation and change. Quite the contrary: because of the regional changes in technology, industry, and communications in the latter half of the century, and because of the inhabitants' personal mobility and other creative responses to these changes, one can argue that the British Caribbean islands in 1900 were populated by much different people than in 1850, a point that probably is as true figuratively as it is literally.[62]

A growing trend saw rural people moving to cities and towns and thereby becoming urban dwellers. Between 1881 and 1891, Castries, St. Lucia, to cite a typical example, grew by more than 2,000 people. Many had migrated from the countryside, a change mildly condemned by local officials as "an accentuated instance of the unhealthy tendency which obtains generally in other lands of crowding into towns at the cost of the country in the way of diminished agricultural production."[63] These demographic changes had cultural implications as well; older people in the region today still recall that early in the twentieth century, individuals who moved from villages to capital towns sometimes took on affected modes of dress and even different speech accents, self-conscious patterns of behavior that almost certainly held true in the late 1800s. These movements from villages to town were not, however, irreversible. Rural peoples sometimes took their pigs and chickens into town with them, giving some areas of the cities a distinctly countrified (and noisy) atmosphere.[64] Socially more important, friends and kin of the new urban dwellers still resided in rural areas, and visits back and forth to celebrate births, funerals, holidays, and the like reinforced social ties between urban and rural zones of individual islands. Nor were these linkages without economic importance. Depressed conditions in the country or a call for laborers in urban areas attracted workers into towns for brief periods for wage earning opportunities. Similarly, wage jobs in rural areas—sugarcane harvesting, road construction, and the laying of water pipes—occasionally brought men into the country from the towns and cities.

The population data arrayed (from south to north) in Table 2.1 do not show the shifts within St. Lucia or elsewhere, yet they are useful in displaying overall island population magnitudes and general trends. These census numbers, further, were no more than decennial snapshots; census officials, especially those in the smaller places, often appended verbal narratives to their reports, explaining that a true picture of the local population was unavailable because so many island residents habitually traveled elsewhere to work. The table is perhaps most useful because it shows human population in a regional context. Trinidad's remarkable population growth, for example, occurred not only because of the influx of indentured Indians late in the 1800s. Much of Trinidad's population increase came at the expense of nearby places such as St. Vincent and Tobago, each of which changed relatively little in human population in the last thirty years of the nineteenth century, in part because so many went to Trinidad. This migratory trend is borne out by Trinidad's census data; persons included as part of Trinidad's population census totals but whose nationalities were recorded as "other British colonies" in the late 1800s were 24,000 in 1881, 33,000 in 1891, and a remarkable 47,000 in 1901.[65]

The interisland migrations of working peoples were qualitatively different from the fluidity of intraisland rural-urban movements in important ways. Although individuals usually decided whether or not to travel, much of the movement to other islands—data for which were usually sketchy and unreliable—often was sponsored and facilitated by colonial officials and planters. Migrants' work usually involved activities such as the seasonal sugarcane harvests in Trinidad and British Guiana or the construction of the dockyards in St. Lucia in the 1880s, locales that thereby became hives of information exchanges among individuals from many different places. Furthermore, these migrations abroad usually were altogether new experiences for migrants from the smaller islands. The novelty of encountering those with different backgrounds, dealing with strange and hostile employers, and being considered outsiders (and therefore responsible for everything from prostitution to petty crimes) helped to account for the sea changes that sometimes transformed migrants' personalities once they were abroad.[66] In most cases, labor migrants returned home to tell those left behind what life was like elsewhere. But some migrants stayed in their new destinations and took up new lives, obviously without the nearby support of the friends and kin they had left behind. Almost certainly the behavior patterns of the large number of recent migrants into Trinidad in the late 1800s, people without recourse to local systems of social and economic support and who therefore had to cope with problems through adaptation or confrontation, must have been a key to understanding why Port of Spain was known as simultaneously exuberant, troublesome, and alive and in many ways an urban center for the entire region.[67]

The complexity and variety in ethnicity, color, and class found throughout the entire Caribbean region were well exemplified among the human populations of the British Lesser Antilles by the end of the nineteenth century.[68] The large majority living there were the black descendants of African slaves. Emancipation a half-century earlier of course meant that these people were free men and women, yet their activities, movements, and livelihoods continued to be dominated by smaller groups of whites and mixed-blood peoples. A handful of Chinese and Portuguese were shopkeepers in the towns and cities, and a relatively small number of indentured men and women from India were agricultural workers in the Windwards, different from the tens of thousands of indentured Indians by this time working in the sugarcane fields in rural Trinidad and British Guiana. The overriding white-black asymmetry in the region, which, as much as anything else, reflected

Table 2.1. Human Populations of the Lesser Antilles, 1861–1901

	1861	1871	1881	1891	1901
Trinidad	84,438	109,638	153,128	200,028	255,148
Tobago	15,410	17,054	18,051	18,353	18,751
Grenada	31,900	37,684	42,403	53,209	63,438
St. Vincent	31,755	35,688	40,548	41,054	47,548
Barbados	152,275	161,594	171,860	182,306	No census
St. Lucia	26,674	31,600	38,551	42,220	50,354
Dominica	25,065	27,178	28,211	26,841	28,894
Montserrat	7,645	8,693	10,083	11,762	12,215
Antigua	36,412	35,157	34,964	36,119	34,971
Nevis	9,822	11,680	11,684	13,087	12,774
St. Kitts	24,440	28,169	29,137	30,876	29,782
Virgin Islands	6,362	6,651	5,287	4,639	4,908

Data are taken from the relevant sections of the British Sessional Papers for 1862, 1863, 1873, 1882, 1883, 1884, 1893–94, 1902, and 1903. Apologies are in order to Anguilla, Barbuda, and the Grenadines, which were occasionally grouped with nearby British colonies but in most years were not mentioned.

the colonial status of each island, could be seen every day in the local cane lands. Where sugarcane plantations dominated the local economies, such as in Antigua, St. Kitts, St. Vincent, and Barbados, white estate managers and white overseers supervised the seasonally intensive toil of black laborers. In Trinidad most plantation field labor was by this time performed by imported Indians, with crews of black men handling most of the industrial work in local grinding mills and boiling houses.

Although white skin and high status usually went together throughout the region, in places where relatively few whites lived, such as in Grenada and Dominica, brown-skinned, mixed-blood men held positions of local power. In Grenada by the late 1880s, prosperous mulatto men dominated rural parish boards and wielded extraordinary influence.[69] Mulattos were even more powerful in Dominica, rivaling whites for local political control; in the 1890s, in the midst of heated debate over the island's political future (a disagreement also involving economic and religious complications and identities), one leading brown-skinned legislator declared a preference to see "Dominica burned to ashes" rather than have the island come under Crown Colony rule and thereby stricter white-imposed control.[70] Whites countered this implied threat with a timely geographical catchphrase, which probably was lost on no one, warning that a few (brown) "demagogues" should not be allowed to transform Dominica into a "second Hayti."[71]

Within the broad constraints imposed by insularity, tropical climates, and British colonialism, the largely rural populace of the eastern Caribbean at the time actually varied considerably from one island to another. Rural peoples in the mainly sugarcane islands, or sugarcane districts of larger islands, usually inhabited village areas consisting of aggregated clusters of tiny houses on rented land plots owned by and contiguous to the plantations. In Dominica, St. Lucia, and Grenada, in contrast, many of the rural villages in the forested highlands and in the isolated windward coastal regions were far removed from the daily surveillance of white powerholders. These latter communities, moreover, were populated by French Creole speakers, and their linguistic identities set them apart from the English-speaking authorities who inhabited the capital towns.[72] A certain village independence also characterized the smallest places, such as Anguilla, the British Virgin Islands, and the Grenadines, where, for better (little government supervision) or worse (an almost total absence of medical care, poor relief, and postal services), communities of seafarers carried on daily lives well outside the direct supervision of colonial officials.

The rural dwellings themselves varied throughout the region as well. The collapsible wooden chattel houses of rural Barbados, for example, differed from some of the older, mud-walled, thatch-roofed houses in the highlands of Grenada. Yet all of these dwellings were, with few exceptions, overcrowded and unsanitary. Probably all of the rural houses of the Lesser Antilles in the late 1800s were qualitatively similar to the average peasant family dwelling of rural Trinidad, where a typical small wooden house had an open front, a single room in back where everyone slept together, and an open cooking fire outside, with washing and bathing in a nearby stream or water hole.[73]

Rural peoples coming in ever greater numbers to reside in the region's towns and cities in the latter decades of the nineteenth century were seeking refuge from periodic downturns in agricultural wage labor and better economic and social opportunities offered in urban areas. But their presence annoyed white officials, who doubtless preferred them out of sight and working in rural areas where they belonged. So unflattering official descriptions of the squalidness and despair of their city housing areas abounded, and these comments in many ways reflected the supposed moral shortcomings of the people who resided therein. For example, in the late 1800s the inhabitants of the small wooden shanties on the outskirts of Castries, St. Lucia, who fashioned their tiny houses from scraps of lumber and metal were creating hazardous conditions because their shacks provided potential fuel for fires.[74] In Basseterre, St. Kitts, in the late 1890s the clusters of closely packed wooden houses in the town were said to lack air circulation or ventilation to relieve the stench from "the numerous cesspools" that marked the poorest sections of town; these physical circumstances, in turn, created "immorality and disease, and, through immorality more disease."[75] Not all of the urban areas of the region displayed similar impoverishment, of course; in the late 1880s some of the new white suburbs of Bridgetown, such as at Hastings and Collymore Rock, created demand for the building products that the newly proposed local brick and cement company could provide.[76]

Nor was disease in the islands confined to the urban housing zones in St. Kitts, St. Lucia, or elsewhere. Individual members of ship crews routinely introduced pathogens to the islands, as they had for many decades (notably during the savage cholera epidemic of the 1850s), and periodic flare-ups of communicable diseases such as measles, influenza, and mumps often followed hurricanes or drought, affecting malnourished urban and rural populations alike. By the 1880s most of the islands had parish medical officers who also administered rudimentary poor relief, and their published quarterly reports detailing illness and mortality provided what probably were fairly accurate depictions of human well-being for the region as a whole. In terms of overall health, the economic depression that began in 1884 hit the sugarcane islands particularly hard because it lowered the wages that were used to purchase imported foodstuffs, leading directly to malnutrition and then disease. These economic, social, and medical circumstances also helped to explain locally high rates of infant mortality, a situation especially prevalent in Barbados.[77]

Official concerns about local health conditions probably had more to do with avoiding economically devastating quarantine sanctions than they did with genuine sentiments about the well-being of local populations. In 1901 and 1902 St. Lucia authorities strenuously denied the apparently obvious local outbreak of yellow fever among the soldiers of the West India Regiment stationed in the barracks above Castries because quarantine, if imposed by neighboring places, would threaten the lucrative ship traffic to and from the island's new coaling station.[78] Similarly, island officials in Barbados fought public acknowledgment of the 1902 smallpox epidemic, yet the port at Bridgetown was quarantined for nearly a year until the island was considered safe. When the Royal Mail Steamers returned in April 1903, the ships were gaily decorated with flags, and "boatmen and lightermen, and others of that class . . . were naturally very glad to see the ships return again to their former headquarters."[79]

Keeping shipping lanes open was important to the region's common people because it allowed them to expand their livelihood opportunities beyond their home islands.[80] The cane harvests in the southern part of the region were by no means the only wage migration destinations. Men from every island traveled away to work, and travel routes were often based on proximity or linguistic identities. The French effort to dig a canal across Panama in the 1880s attracted black men from throughout the region, notably French Creole–speaking St. Lucians. Similarly, Dominicans commonly traveled to the gold placers in the interior of French Guiana. Laborers from St. Kitts and Antigua went to work at the nearby Danish possessions of St. Thomas and St. Croix. Though their identities and destinations varied, traveling men invariably were indicted for countless sins, ranging from creating trouble abroad to leaving women and children behind. Early in 1885 immigrant Barbadians were said to have introduced "lewdness and immorality" to the formerly "innocent amusements of the Carnival" in Trinidad.[81] A decade later widespread burglary in Roseau, Dominica, was blamed not on "honest and trustworthy" Dominicans but on the "men from other islands, principally Guadeloupe and Martinique."[82]

Unflattering stereotypes aside, these migrants' experiences both at home and elsewhere doubtless provided a reservoir of firsthand comparative information about the eastern Caribbean, information that the migrants must have disseminated routinely among family and friends and anyone else willing to listen to stories of their exploits. Probably better than anyone else, these travelers could rank informally, for example, towns and cities throughout the region on the basis of physical amenities, excitement, and social and economic opportunity. At the low end of the continuum doubtless were places such as Roseau, Dominica, a town of about 7,000 in 1900, with pathways of cobblestone, nearby roads unworthy of wheeled traffic, lighting provided only by a few kerosene lamps after dark, and public buildings that were "in a wretched condition and devoid of paint."[83] At the other end, Port of Spain, Trinidad, was the bright, bustling, unquestioned metropolis of the region.[84] The city's streets were nearly all macadam or asphalt, street lighting was provided by the new Trinidad Electric Light and Power Company, and its culturally varied populace of more than 54,000 at the turn of the twentieth century—including all those from neighboring islands who brought with them a medley of different languages, odd-sounding accents, and different styles of clothing—jostled and tolerated and accommodated one another on the city's streets every day, representing altogether "a collage of peoples and cultures."[85]

A variety of mercantile houses, tram service, first-class hotels, public buildings (including hospitals and libraries), and parks made Port of Spain attractive as a regional shopping destination as well as a temporary residence for travelers and officials at the higher end of the social spectrum. In 1887 Moses Sawyer, the U.S. consul in Trinidad, wrote home to describe a place as sophisticated as any he knew about: "There are more linguists-scholars and cultivated people at Port of Spain than anywhere this side of the capitals of Europe," enthused Sawyer, who went on to describe Port of Spain as "the capital of a haughty aristocratic military Crown Colony. At Government House there are gay festivities never known at such cities as Liverpool or Antwerp."[86] The overall cultural diversity on display in Port of Spain's streets and alleyways had its physical-structural counterpart in the startling variety in housing patterns within the city. From practically any vantage point, one could see houses and apartments of the middle class and well-to-do juxtaposed with the hastily thrown-up "barrack yard" sheds that housed the recent poor immigrants into the city, who came from rural Trinidad and elsewhere. Among other things, these housing patterns meant that members of widely varying social classes were rarely separated. The gardeners, the day laborers, and especially the maids and nannies who represented "often the predominant influence in the lives of the children" of Trinidad's upper classes resided, more often than not, in the barracks scattered throughout the city.[87] This proximate familiarity is another reason Port of Spain was a locus of countless altercations, including political protests and demonstrations, in the latter half of the nineteenth century. Reports of these goings-on filled the newspapers of the region, not only in Trinidad, and details of such incidents were doubtless retold by travelers when they went back home to St. Vincent, Martinique, or Barbados.

Within the city the Government House, or what came to be known as the Red House, situated opposite Brunswick Square, was the usual focal point of political and social protest. As early as October 1, 1849, rock- and debris-throwing occurred outside the Government House, where the people's representatives were complaining to members of the local colonial government about the treatment of imprisoned black debtors.[88] The following day and, significantly, in the years thereafter, similar protest marches naturally gravitated toward the Red House, with police—and occasionally members of the West India Regiment stationed in Trinidad—summoned to provide protection. The Red House thereby came to represent a cumulative architectural symbol of the colony's (and Port of Spain's) importance as the century unfolded. All the while, builders and architects enlarged and contributed to the edifice, and its completion in mid-1897 was planned to coincide with Queen Victoria's diamond jubilee. In commemorating this event, invited dignitaries and others viewed with pleasure "the imposing and ornamental pile of buildings so well situated opposite Brunswick Square."[89]

Social discontent marked by public protest in Trinidad, as well as in the whole colonial Caribbean, appears to have been inevitable. In a repressive local atmosphere where locals knew about increasing political equality, relaxation of the voting franchise, and the formation of labor unions elsewhere in the world, it seemed on occasion that emancipation had never come about. With few exceptions, the region's black people had no say whatever about the social, political, and economic conventions under which they lived. The resulting social disturbances, such as the Morant Bay rebellion in Jamaica in 1865 and the Confederation Riots in Barbados in 1876, recently had led to the imposition by the London Colonial Office of direct Crown Colony government for most places in the eastern Caribbean, a move generally interpreted "as a means of protecting the interests of European planterdom."[90] In a broad sense, Crown Colony government—which lessened the influence of even white-dominated elected legislatures—was simply a new twist in a seemingly endless struggle, because the attempt to achieve political parity in the region was, by the late 1800s, decades old. A sympathetic tribute to one Isaac St. Orde (born in 1827) appearing in the *Dominica Guardian* in June 1904 made the point: "He participated in all the struggles resulting from the determination of the Colonial Office to deprive the coloured and black sections of the population of this and other West Indian islands of citizenship under the British Empire."[91]

From London's point of view, the differing and changing political arrangements developed by the British Colonial Office to govern the individual island colonies (and multi-island groupings) of the British Lesser Antilles represented confusion and an ignoble concession to the petty peculiarities of insular political geography. When the royal commissioners traveled through the region in 1897, they visited the colonies of the Leeward Islands (composed of the separate "presidencies" of Dominica, Montserrat, Antigua, St. Kitts–Nevis, and the Virgin Islands); the Windward Islands (St. Lucia, St. Vincent, and Grenada); Barbados; and Trinidad-Tobago. These particular groupings, however, represented little more than moments in time. Tobago and Barbados, for example, had been grouped for governance purposes with the Windwards from 1833 to 1885 and ruled by a British governor in chief. Beginning in the latter year a single governor had jurisdiction in Barbados, and four years later, in 1889, Tobago became attached administratively to Trinidad.

Although all of these colonies were ultimately ruled by the Crown, the customary means by which local politics were carried out varied widely from place to place. Sir Francis Fleming, governor of the Leeward Islands, complained to the 1897 commissioners about the bewildering variation in how, for example, meetings were conducted and taxes imposed from one island to the next—even within the same colony—and he suggested some sort of standardization or political confederation as possible remedies. Dominica and Antigua, for instance, were each ruled by a grouping of elected (by very few voters) and appointed officials; yet in the former island a local administrator presided over the legislators' meetings, and in the latter he did not. Neither island had a system such as that in Barbados, where a planter-dominated assembly wielded unusual power. Nor did mixed signals from London, according to Fleming, help in local governance, because this higher authority had the unnerving propensity to, for example, "give the different Presidencies to understand that they are to be governed by . . . central authority one day, and to appoint local officers with increased jurisdiction the next."[92]

The working peoples had reasons to interpret the overall political structure in less complex ways. What the British perceived as perplexing variety, local laborers experienced as monolithic oppression. As only one example of the harassment and intimidation occurring throughout the region, Police Inspector Wright in Grenada in 1885 tore down houses of those he disliked, a whimsical and capricious enforcement of a ten-year-old law in St. George's prohibiting the erection and repair of wooden houses.[93] Wright and probably a majority of local police magistrates were white, whereas the police constables who patrolled towns and villages every day were black. In most cases police constables were either from "foreign" islands or other parts of the same island so as to lessen the possibilities of solidarity between the policemen and the policed. Police intimidation, combined with hazy, ill-defined, and ever changing sets of rules and local laws, also fueled innumerable minor altercations that intensified the ill will between people and police. And in a probable majority of cases, fines and arrests were carried out when simple warnings would do. Marketplaces throughout the islands were animated by continuous cat-and-mouse contests between police constables and vendors who had failed to purchase locally prescribed "licenses" to sell their goods. In 1896 in St. Philip parish in Antigua, thirty-five black villagers were fined ten shillings apiece (a formidable sum for pitifully paid estate workers) for disturbing the peace by "loud drumming." The fines were exacted by estate managers and magistrates, who were at times the same person. Furthermore, the punishment was not preceded by warnings, and most of those fined were ignorant of the rules they had broken.[94]

Police harassment was more a function of the reigning political regimes than it was a manifestation of the inherent evil of law enforcement personnel. Much of the animosity between black townspeople and villagers and local policemen, for instance, centered on the efforts of the latter to collect taxes, especially in rural areas where other forms of public "service" were nearly nonexistent. In general, the British Colonial Office expected individual colonies to pay most of their officials from locally extracted taxes. So the insensitivity that London often exhibited toward the wide variations in individual island cultures and economies was very different from the specific taxation schemes British administrators concocted to fit particular local circumstances. Where sedentary sugarcane workers depended on wage-purchased food imports to stay alive, such as in Barbados, St. Kitts, and St. Vincent, import duties on flour, butter, and corn (which meant higher prices for destitute villagers) provided the bulk of local tax revenues. On the other hand, where country people were more self-sufficient for their subsistence, the police collected (or tried to collect) the loathsome head taxes or "road taxes" from local villagers, and confiscation procedures sparked endless wrangling. The Road Act of 1880 in Dominica called for head taxes for every adult male to be paid in January, May, and September, with defaulters liable to one month's imprisonment with hard labor. Rural taxpayers were warned that this new tax scheme would be enforced vigorously and that the "dilly-dallying" tax-avoidance ploys of the past would no longer be tolerated. In January 1882 local police traveled from one Dominican village to another ringing a bell to signal that taxes were now payable. But the severity of these taxation schemes in the Dominican countryside and elsewhere was far greater than their humorous imagery. In 1890 eighty-year-old Paul Remy of Prince Rupert's Quarter in Dominica was imprisoned because of his inability to pay the road tax, and he subsequently died in jail. Remy's death was locally condemned as an outrage, commentary doubtless interpreted in London as radical ranting from an island where men of color had achieved undue influence and power.[95]

Despite the widespread regional resentment of the taxation policies, institutional racism, and political oppression in the Lesser Antilles at the end of the nineteenth century, it would be wrong to suggest a universal dislike and mistrust of everything British or European. A small but growing and very influential stratum of black and brown West Indians had received formal educations and held professional positions, and they were steeped in British law, literature, and the arts. Increasingly literate men and women in working-class positions, furthermore, read books and newspapers, were faithful members of Christian church denominations, and made it their business to acquaint themselves with events both at home and abroad. They were not, they often exclaimed to one another, the heathen African or Asian "natives" that the British were encountering elsewhere in the empire. A notion of this complex, multifaceted, and important cultural characteristic of the British Caribbean at the turn of the century was perhaps best captured in the context of particular incidents. At Queen Victoria's diamond jubilee in London in 1897, she encountered in a receiving line the black Trinidadian lawyer and reformer Emmanuel Lazare. The curious queen asked what language people spoke in Trinidad. Lazare replied, "Madam, in Trinidad we are all English."[96]

Nor was every British administrator universally condemned and disliked. West Indian writers and spokespersons openly admired the generosity and sympathy articulated by colonial administrators such as C. S. Salmon and Sydney Olivier, the latter a Fabian Socialist, secretary of the 1897 Royal Commission, and a future governor of Jamaica. Similarly, the radical white jurist Sir John Gorrie of Trinidad and Tobago maintained a "vigilant, fearless, and painstaking" insistence that the working peoples of the region and the empire were due full equality and justice under the law, and he was widely admired among members of the local working class.[97]

But officials of the quality of Salmon, Olivier, and Gorrie were the exceptions. In most other cases local peoples encountered patronizing and ineffectual white administrators of only medium competence whose sole advantage over locals was the color of their skin.[98] The resulting resentment of these officials among educated locals, who had genuine stakes in their own societies, was therefore distinct and different from the resistance exhibited by black workers on sugarcane plantations. Among the most articulate intellectuals of the region at the time, J. J. Thomas of Trinidad, a self-taught black linguist, excoriated the patronizing, racist hypocrisy of British Caribbean colonial rule, personified by "highly salaried gentry . . . who have delighted in showing themselves off as the unquestionable masters of those who supply them with the pay that gives them the livelihood and position they so ungratefully requite."[99] Even more important than expressing disgust or condemnation, educated West Indians formed an intellectual elite whose leadership and organizational skills—combined with the muscle that restless, sometimes torch-wielding crowds could provide—were effective in combating the negative excesses of British colonialism.

At the positive end of the spectrum of social and political sentiment in the eastern Caribbean region at the turn of the century, the good things about being "English," probably all would agree, often were focused on the embodiment of Queen Victoria (despite her obvious naïveté about a small island colony in one corner of her empire). The queen had reigned forever, it seemed. Her ascendance to the throne in 1837 was nearly simultaneous with slave emancipation itself, and everyone in the islands, rich or poor, young or old, celebrated the queen's birthday each June with gusto. June 22, 1897, therefore represented a special date, not only for British West Indians but for all British subjects because it marked the sixtieth anniversary, or diamond jubilee, of Victoria's reign. Amidst pageantry and merriment, the event was celebrated throughout the empire, including in all of the small islands of the West Indies, where the celebration involved, among other things, the lighting of fires.

3

Fires in Towns and Cities

ON EASTER SUNDAY 1897 AT THE CARLTON CLUB in London, the author and politician George Baden-Powell composed what would soon become a widely distributed letter of appeal. His letter, eventually published in all of the major newspapers of the British Caribbean and presumably throughout the rest of the empire, forwarded a novel suggestion. Baden-Powell pointed out that the London committee arranging the festivities to mark the sixtieth anniversary of Queen Victoria's reign on June 22, 1897, had proposed an arresting idea. At 10:00 P.M. on the evening of the twenty-second, "beacon bonfires" were to be lit on hilltops throughout the length and breadth of the United Kingdom, altogether creating a gleaming tribute to Victoria's benevolent reign, a collective and fiery celebration involving thousands of participants and onlookers and one that would extend from Dover north to Inverness and from Yorkshire to beyond the Irish Sea.

"Cannot this bright idea expand to the oversea provinces and out-posts of the British Empire?" Baden-Powell queried. "Cannot the Great Signal blaze from the thousand historic heights of the Empire . . . from Table Mountain and the Metopos; from Mount Macedon and the Blue Mountains; from Adam's Peak and Simla . . . and the rest?" He proposed that bonfire committees in every British colony could be formed easily, following the "rapid currency" the press would give his written suggestion. Eventually, according to Baden-Powell, the many simultaneous fires everywhere on that night would form a "grand symbol" throughout the world for all persons involved in "lightning up a Fiery Cross to signalise peace, prosperity and progress for all who enjoy and profit by the beneficent sway of Victoria the Great."[1]

Baden-Powell, who fifteen years earlier had acted as a visiting royal commissioner to assess the finances and administrations of Jamaica and the Windwards, would have been pleased at the alacrity with which British West Indian subjects warmed to his suggestion. On June 22 Antigua and Jamaica commemorated Victoria's diamond jubilee with evening fireworks. Barbadians held an illuminated bicycle procession. A candlelight parade proceeded from Cayon Village, on St. Kitts's windward coast, all the way southward to Basseterre. A torchlight parade in Roseau, Dominica, was followed by coastal bonfires in nearby communities. Vincentians and Grenadians set bonfires on the prominent hills above their capital towns of Kingstown and St. George's. "Evening illuminations" greeted onlookers on the streets of Port of Spain, Trinidad, and Georgetown, British Guiana, although rainy weather marred the celebrations in both of these southern Caribbean cities.[2]

Yet it must have been at the village level in the West Indies, where everyone participated, knew one another, and became immediately involved,

Port of Spain, Trinidad, decorated for Victoria's diamond jubilee, 1897 (from Gerard A. Besson, *The Angostura Historical Digest of Trinidad and Tobago* [Trinidad: Paria, 2001])

where the fire-lighting ceremonies were the most meaningful. In the small fishing village of Soufrière, on the leeward coast of St. Lucia, the bells of the Roman Catholic church pealed at midnight on the twenty-second. At that instant men lit bonfires along the water's edge that illuminated the nearby fishing boats and wooden houses, which cast their elongated shadows onto the green hills above to produce "an appearance most magnificent." A crowd of 3,000 villagers had turned out, and their prolonged cheers of "Vive la Reine!" carried over the water and "reverberated over the hills and valleys." The small town's fires burned for more than an hour. After Victoria's blazes died out, the distant sound of drums signaled the entry into Soufrière of dancers and torchbearers from nearby hamlets who then extended the celebrations far into the night.[3]

Had the 1897 British Caribbean fire celebrations been mandated through official government channels, it is likely that local enthusiasm for them would have been greatly reduced. To be sure, the revelry and tribute in Soufrière and the other towns and villages of the British Caribbean paid homage to the sovereign and ultimate symbol of the British Empire. Yet lighting fires in honor of distant and idealized Victoria bore little relation to the unpleasant daily encounters between Caribbean peoples and local colonial officials and, more often, with the magistrates and police constables who enforced the local laws. It is a near-certainty that as the revelry progressed at places like Soufrière village during the nighttime hours of June 22, 1897, the dancing, drumming, drinking, and general merriment took directions that George Baden-Powell never had in mind and that had little to do with Victoria's diamond jubilee.

Just as certain must have been the sense of unease with which the local colonial officials anticipated the fiery celebrations. Fireworks displays were nothing new in the region. By the late 1800s, public fireworks marked holidays and anniversaries in many places in the world, and a variety of colors and increasingly elaborate designs, patterns, and artistry made these public events widely anticipated.[4] Victoria's golden jubilee ten years earlier, for example, had been marked in Port of Spain, Trinidad, by evening fireworks at the Queen's Park that drew a large crowd from among all segments of the city's population. The brilliant, multicolored display, imported from London, featured, among other things, a sparkling likeness of the queen herself. Even more momentous on that night had been the projected beams of an electric searchlight from the HMS *Comus* onto Port of Spain's harbor complex and city buildings after the fireworks were spent.[5]

Yet general celebrations featuring torches, open flames, and bonfires were far more dangerous and provocative than were fireworks in public parks. Under normal circumstances, and even for special occasions, British

Caribbean colonial authorities usually discouraged or prohibited displays involving fire. The plans for torchlight parades marking the fiftieth anniversary of full slave emancipation in August 1888 had been openly opposed by merchants, planters, and some of the government functionaries in Trinidad and were only grudgingly accepted elsewhere in the region. The showery weather throughout the eastern Caribbean that accompanied that particular date had played a fortuitous role, as far as government officials were concerned, in helping to curtail and subdue the festivities that might otherwise have gotten out of hand.[6]

Fire, especially within the confines of very small islands, obviously attracted crowds, and it also always seemed to activate them. Furthermore, the destructive potential possessed by groups of people was well understood by the white planters and government officials in the British Caribbean in the late 1800s. In an era when European statesmen and scholars had begun to discover the dangerous importance of urban masses, crowds, and "mobs" that accompanied a new industrial order on the Continent, Caribbean officials already had been wary of groups of their own laborers for decades. During slavery, when blacks on some of the islands had been allowed to participate in seasonal revelries that involved dancing and singing, "guards" of whites had stood by lest trouble occur. In the years after emancipation, parades and demonstrations had been similarly monitored in case demonstrations became destructive outbreaks, disturbances or riots, or heaven forbid, islandwide "risings."[7]

The demographic and settlement characteristics of most British Caribbean sugarcane plantation islands at the end of the nineteenth century reduced the potential for the coalescence of truly large crowds and associated group demonstrations, with or without fire. In most cases the workers' settlements were scattered throughout Caribbean countrysides, and travel from them was nearly always accomplished on foot. But topography, landforms, and general accessibility varied from one island to the next. On reasonably flat and internally accessible Barbados, market women and others walked as many as ten miles from countryside to town and back again, often two or three times in a single week. Similarly, on the fertile and gently sloping volcanic soils north of Basseterre, St. Kitts, sugarcane estate workers' villages were in easy walking distance of town. If pressing social issues seemed to call for it, rural peoples elsewhere could unite in a lengthy march to the capital town and seat of government there to air grievances. So, for example, in November 1891 when an estimated 2,000 men from the St. Vincent countryside traversed the winding coastal roads bordering the island's precipitous volcanic interior to come to Kingstown to protest rumored political changes that they thought would end their emigration prospects, local authorities telegraphed the HMS *Buzzard* and its contingent of British bluejackets for protection.[8]

Several of the men from the St. Vincent countryside in November 1891 had carried lighted torches into Kingstown, a clear and well-understood sign of real trouble. Beyond the mere sizes of groups or crowds, fire or the threat of fire, in St. Vincent and elsewhere at the time, often signaled an extraordinary level of distress. When the activities of crowds, parades, or demonstrators took turns for the worse in the islands at the turn of the century, fire usually accompanied and even seemed to inflame the attitudes of those involved. This rule (or at least this implicitly understood observation) alerted and simultaneously produced anxiety among British West Indian planters and colonial administrators alike. This anxiety manifested itself because Caribbean fire at the turn of the century, beyond its obvious destructive potential, often delineated the cleavages in local societies. Working peoples used fire, among other things, to signal discontent, while powerholders pushed hard in the other direction to control it.

The principal means by which officials attempted to control fire was to stress its destructive capacity and thereby devise legal restrictions against it. By the late 1890s every island had laws prohibiting and regulating fire in both urban and rural areas, rules regularly discussed, revised, expanded, and modified by colonial legislative bodies. St. Kitts and Barbados prohibited the growing of canes too close to buildings. Anyone carrying "any lighted torch or other matter in a state of ignition" in Trinidad could be fined as much as £5. Government-issued licenses were necessary for setting fires to clear brush in most places. In the northern islands, fire prevention laws took into account seasonal drought that desiccated both plants and wooden buildings. A Jamaican planter at the turn of the century asserted that he knew "of no place where forest and other fires are more prevalent than in Jamaica" and concluded "that the only way to stop them is by legislation prohibiting their use or allowing them only under certain conditions."[9] Yet representatives from any of the other islands of the British Caribbean would have contested the notion that Jamaica was uniquely afflicted by fire, pointing out that their islands, too, had the same kinds of fire problems and antifire laws as Jamaica did.

Flashpoints could and did result where (authorities') antiburning rules opposed (peoples') fire practices, and these points of conflict were most noticeable in the capital cities and towns where both the seats of government and main urban populations resided. Such an incident occurred at St. George's, Grenada, late in 1885. The town's merchants recently had convinced a new governor that prohibiting the usual fire celebrations that marked Guy Fawkes night would protect the wooden buildings on the capital town's market square. For years townspeople in St. George's had celebrated the Gunpowder Plot by flinging flaming, pitch-soaked projectiles about the square on the night of November 5. Consequently, the ill-advised prohibition of these celebrations for 1885 inspired resistance from the people, who naturally felt that they had been denied a customary right. The form their resistance took was accented by fire itself, as groups of "rowdies" ignited tar barrels at key points around the town on the several days leading up to Guy Fawkes night before actively confronting local police and their allies with brickbats and stones and torches on the night of November 5.[10]

Cannes Brulées

By the 1880s and 1890s in the eastern Caribbean, conflicts between police and peoples that were centered on the curtailment of traditional rights involving fire celebrations were anything but new. Indeed, the only novel features of the 1885 Guy Fawkes disturbances in Grenada were the island's identity and the season of the year. That is because in Trinidad for more than a generation, a running battle—involving propaganda, rival strategies, skirmishes, and occasionally violent confrontations—between island authorities and local, torch-wielding celebrants had been fought over the issue of "Camboulay," or *cannes brulées*, French for "burning canes." These pre-Lenten torchlight celebrations (widely acknowledged as the forerunners of Trinidad's famous annual Carnival celebrations) also were known as "canboulay" or "canne boulee." Yet however they were pronounced or spelled, there was little ambivalence over the importance and the contested character of Trinidad's torchlight parades, which early in the 1880s and thereafter annually pitted the authorities' sober control tactics against the exuberance and organizational skills of the participants, resulting in confrontations focused precisely on the control and use of fire.

Since the 1860s in Port of Spain, the pre-Lenten carnival celebration had been dominated by the fiery *cannes brulées* processions. The torchlight parades of (usually) masked participants began at midnight on Carnival Sunday and lasted for three nights running, ending at dawn on Wednesday. The processions featured not only open flames but also drumming, shouts, bawdy songs, and boisterous, noisy confrontations and even fights between rival groups or bands. The participants were mainly black Trinidadians from the "lower orders" of the populace, people who inhabited Port of Spain's "barracks" and "yards" that had proliferated to accommodate the city's expanding population in the years after emancipation. Some assemblages of black celebrants were composed of recent immigrants, adding a competitive element to the activities of different torch-wielding groups. At the other end of the social spectrum, many if not most whites doubtless considered the *cannes brulées* parades excessive, offensive, and perhaps most important, frightening exhibitions of the licentiousness and lawlessness and rebellious violence they routinely associated with the city's black underclass. The size and complexity of Port of Spain's population by the turn of the century, different from any other island in the region, contributed to the unease and difficulty with which white officials tried to maintain order in the weeks preceding the torch parades. Many black Trinidadians understood English but conversed among themselves in Creole, thereby distancing and alienating themselves even further from the authorities. White officials countered with mobilized squads of English-speaking police from Barbados, thereby guaranteeing mutual and widespread antagonism.[11]

In early February 1880, official attempts to reign in the excesses of the Port of Spain carnival celebrations—and indirectly to exert greater authority over the city's "lawless" class—manifested themselves in specific actions taken to control fire. Citing an 1868 law prohibiting the possession or use of any torch that could be interpreted as an "annoyance or danger" to others, Captain A. W. Baker, the city's new police commander, confiscated the torches carried by pre-Lenten revelers. The following year, in 1881, heeding warnings and threats of trouble emanating from several of the black housing areas, Baker armed the police with clubs to carry out the antitorch decree; the result was the Carnival Riots, in which street lamps were smashed and 38 of 150 armed police were injured. In the aftermath the revelers held a mocking torchlight parade while jeering Captain Baker. The island's governor, Sir Sanford Freeling, insisted that his antitorch decree had had nothing to do with an attempt to "interfere with the amusements" of the city's residents but, rather, was intended to protect Port of Spain's buildings from fire in the dry season. Freeling added his surprise that the riots had been so intense: "I . . . had no idea that they attached such importance to their 'Camboulay.'"[12]

Doubtless the understanding of the importance and meanings of the pre-Lenten Camboulay celebrations varied greatly, not only among observers like Freeling but also among the participants themselves. In attempting to seek answers and to provide background to the recent events, the *Port of Spain Gazette* published, in both French and English, an article titled "The Origin of Canne Boulee" in late March 1881. The piece stressed custom and continuity, pointing out that the annual event's roots went well back into the island's slave era, when groups (bandes) of slaves from different estates marched, accompanied by torches and drums, to assist in harvests or "when a fire occurred on the plantation." The tradition, the article went on to explain, continued after emancipation, only now rival bandes from different neighborhoods from Port of Spain often confronted one another with stick-fights or other competitions, and the celebrations had "become more boisterous and noisy than heretofore."[13] Fire, history, and realizations of freedom itself thereby all converged in the *cannes brulées* processions of the late 1800s. In order to understand these issues better, the commissioner who came from London to Trinidad to investigate the causes and extent of the 1881 riots, Sir R. G. C. Hamilton, actively sought advice as to the significance of the fire celebrations. A local historian confirmed to him the slavery origins and added, "After emancipation the negros began to represent this scene as a kind of commemoration of the change in their condition, and the procession . . . used to take place on the night of the 1st August, the date of their emancipation, and was kept up much for the same reason as the John Canoe dance in Jamaica. After a time the day was changed, and for many years past the Carnival days have been inaugurated by the 'Cannes Brulées.' "[14]

Although best known as an urban event in the late 1800s, *cannes brulées* also was practiced throughout rural Trinidad, where without constriction by a grid of city streets or the officious bullying and curtailment of Port of Spain police, it may have been an even more faithful replica of earlier times. The annual Port of Spain fire celebrations also likely symbolized familiar events for rural peoples from Trinidad and even from nearby islands who recently had moved to the city, representing activities similar to those they already knew from home and that thereby provided an element of ceremonial continuity in their otherwise changing geographical circumstances. For migrants to the city from places where the torchlight celebrations were unknown, the *cannes brulées* processions can be interpreted as group activities that initiated or acclimatized them into the customs and expectations—not to mention the excitement—associated with urban life.

Country versions of *cannes brulées* in Trinidad were both fiery and frightening. Father Massé, a Catholic priest at tiny La Brea, in the south of the island, found the local pre-Lenten fire celebrations in 1882 far more horrifying than culturally edifying. Awakened at midnight on February 20 by shouts and horns, Father Massé described the subsequent lighting of torches all around the village as representing the cane fires set earlier in the island's history by disgruntled slaves. After that the participants, some armed with sticks, rushed with a frenzy to a central point to simulate extinguishing the fire. Real blood was spilled in the melee. Hideously masked performers then did abominable dances and uttered "the cries of . . . beasts of prey." Confusion, not ceremony, seemed to reign. Many participants carried sharp knives and blunt instruments, a few of the many elements that, taken altogether, created a "spectacle both frightening and truly diabolic."[15] Doubtless the priest saw, in the noisy chaos of the celebrations of *cannes brulées*, the very antithesis of all that he believed and stood for. A more recent academic interpretation of Carnival's origins in Port of Spain reaches the same conclusion. John Cowley, in assessing the overall significance of the *cannes brulées* ceremonies of late 1800s Trinidad, emphasizes that bands of both men and women participated in processions designed to frighten onlookers. These spectacles included "masques depicting death, and demons (in which tar was used to cover the body). . . . All were unified by the use of flambeaux to simulate . . . a cane fire and, perhaps, create an impression of 'hell.' "[16]

In several ways Trinidad's pre-Lenten revelries at the end of the nineteenth century reflected a regional, not simply a local, fire celebration culture. Grenada, St. Lucia, and Dominica, nominally British but with French Catholic cultural traditions, all had the pre-Lenten *cannes brulées* torch parades. Indeed, after the 1881 disturbances in Port of Spain, Commissioner Hamilton pointed out that many of the local Carnival's "most objectionable features" had been brought by the ongoing stream of immigrants from nearby islands. There was even an intracity spatial dimension to the scuffles and altercations between different black groups during the celebrations. Henry Street represented a north-south dividing line between French Creole speakers who resided east of the line and English speakers who lived to the west.[17] Keeping these rivalries in mind, as well as the difficulty of controlling them after dark, Hamilton thought that if the local Carnival processions were restricted to the daylight hours as they were "in the French islands" (he seems to have meant Guadeloupe and Martinique), the entire torch-carrying problem might be solved.[18]

Cannes brulées processions also were lusty events in the Windwards at the time. The torch ceremonies in Grenville, Grenada, though far smaller in scale than those in Port of Spain, reflected not only cultural similarities with Trinidad but also the time-honored Caribbean tradition of blaming immigrant outsiders for local problems. When police constables in Grenville took a disorderly drunk to jail during the torch parade in early February 1894, a group of his friends, said to represent the "rowdy element" recently arrived from St. Vincent, stormed the Grenville station house and broke windows. Local commentary denounced the incident in particular and *cannes brulées* in general as "senseless and immoral" because they provided little more than opportunities for the "worst element of the community . . . to . . . make night hideous by their unseemly noises and howlings."[19] Disturbing noises and ghastly displays similarly accompanied *cannes brulées* in Roseau, Dominica, and alarmed annually the few white officials on the island. In 1892 the fiery parade included a crowd of young people "almost in a state of complete nudity" whose bodies, smeared with molasses and ashes, inspired observers to trace these hellish abominations to "Darkest Africa." The torchlight ceremonies there on February 16, 1893, saw "unruly spirits" coalesce into "an infuriated mob" that threw stones and other missiles at officials and innocent bystanders alike after police tried to rein in the festivities. The similar "obscene, licentious" goings-on the next year in Roseau were interpreted as a custom that took place "only in French or quasi-French communities . . . such as St. Lucia and Trinidad." Finally, the Dominica government's Act #1 of 1895 prohibited, with apparent success, open torches and other excesses of the annual *cannes brulées* celebrations there.[20]

The February fire celebrations nevertheless lasted longer in the smaller islands to the north than they did in Trinidad. By the mid-1890s in Trinidad, the flames had been taken out, through legislation and police vigilance, of the *cannes brulées*, and the term "Carnival" was used more and more for the pre-Lenten celebration. The strict legal prohibition of "the carrying of any lighted torch" was specified by Trinidad's Torch Ordinance of 1884.[21] Despite scattered displays with open flames thereafter, the antifire law was generally effective. But the annual Trinidad celebrations, whose origins were in torch processions, were being transformed, not eliminated. Celebrants continued to indulge in activities designed to offend and shock, whether they be flour throwing, cross-dressing, or obscenity shouting, all of which were roundly condemned each year. The apparent eradication of the fire threat, however, had calmed official nerves. A comment in a Port of Spain newspaper in February 1896 bordered on patronizing, yet it demonstrated how much the custom seems to have been tamed: "The poorer classes have scarcely any amusements, and we think that so long as there is no obscenity or disgraceful conduct they might be allowed to indulge in their long-established custom."[22] Members of the local elite, especially those who could recall the torches of years earlier, were nonetheless wary at the beginning of each calendar year.

"A Terrible Calamity—A Gigantic Conflagration"

Authorities' condemnations of torch parades were not only verbal and legal attempts at social control, but they also reflected the genuine fears of fire damage, especially in urban areas. Most large city buildings in the British Caribbean in the late 1800s were wooden, although recent building codes in most places specified stone or masonry for new construction. Especially in the dry seasons, the aggregations of sun-bleached wooden buildings (or buildings constructed partly of wooden components) represented highly combustible fire hazards. This omnipresent danger was compounded by the innumerable small wooden dwellings—and their inhabitants' kerosene lamps and trash heaps—that crowded alleyways and formerly empty urban land areas, now the dwelling places for those who had recently abandoned the depression-stricken rural areas. In coming to town, these refugees from the economically depressed countrysides had, literally, brought with them more fuel for possible city fires.

The racket of hammers and saws throughout Port of Spain in 1897 continued to ring out from the citywide rebuilding effort following the holocaust of March 1895, an enormous fire characterized by a local newspaper as "a terrible calamity—a gigantic conflagration."[23] Yet these words understate the extent of the fire and the damage that it produced while, amazingly, killing no one. It started on the afternoon of Monday, March 4, probably in Davidson and Todd's store on Henry Street. Fanned by a stiff easterly breeze, the fire enlarged, intensified, and then roared west to Frederick Street, wiping out the city's central commercial district. By nightfall, individual flames of the conflagration were clearly visible from the hill behind San Fernando, twenty miles to the south, to onlookers there who had been alerted by telephone of the catastrophe under way in the capital city. The fire itself was not completely extinguished for four days; during that time it completely destroyed fifty-four buildings of Port of Spain's central business district, created immediate and widespread unemployment, and caused overall losses of about £350,000.[24]

The initial blaze leading to the larger fire had gone unnoticed longer than normal for a Monday because the city center was nearly abandoned that afternoon. A cricket match between a visiting English team and the "All Trinidad" side had attracted a large crowd to the savanna at the city's northern edge, and most merchants had given their employees a half-day holiday to attend.[25] About 4:30 P.M. everyone saw the "billows of black smoke" and rushed south into the city to see the fire. The fire brigade arrived late (their tardiness was a subject of recrimination and discussion in the fire's aftermath), and the city's water supply system failed to live up to expectations. Despite these problems, and damaging as the giant fire was, it would have been far worse ("ten times worse," in the opinion of one observer) except for the heroics of the men aboard the four naval vessels that happened to be in the Port of Spain harbor. About 5:00 P.M., when the wind was beginning to shift to the west and the fire therefore started to turn in on itself, the bluejackets from the HMS *Buzzard* came ashore to help. An hour later the U.S. sailors and marines from the visiting American cruisers *New York*, *Cincinnati*, and *Raleigh* arrived. During the night the American and British crews together fought the fire without respite but with notable success, not only by confronting the huge blaze directly with buckets and hoses but also by tearing down and dynamiting buildings in its path. Several men from the *Raleigh* suffered painful burns from a falling balcony. The armed American

The U.S. cruiser *Raleigh*, part of the task force that helped fight the 1895 Port of Spain, Trinidad, fire (from *Uncle Sam's Navy* [Philadelphia: Historical Publishing Co., 1898])

marines also assisted in crowd control and later helped to prevent looting. The Americans withdrew the following day amidst the effusive thanks of the Trinidad authorities.[26]

The experience of March 1895 reminded all Trinidadians, particularly those living in the capital, of the great fires in Port of Spain's past as well as the city's continuing vulnerability to burning. In August 1896 a good-sized blaze on Frederick Street rekindled fears from the preceding year, but it came under the quick control of the fire brigade and, owing in part to a lack of wind, caused only about £10,000 in property damage.[27] In a broader regional sense, the 1895 Port of Spain catastrophe became one of the Caribbean's great city fire benchmarks of the late nineteenth century; news and recollections of them, like hurricanes and earthquakes, touched everyone in special ways. These fiery calamities affected not only those with direct experience but also others who learned of the events from friends and family or who read about them in their local newspapers. Much of the commercial area of Kingston, Jamaica, burned down in December 1882 when a sea breeze pushed the fire inland ("leaping from roof to roof") in the afternoon and then the evening land breeze drove the conflagration downhill to destroy the remaining wharves and warehouses at the waterside.[28] The terrible Martinican fire of June 1890, perhaps ignited by an unattended coal pot or kerosene lamp and then spread by windy weather conditions, killed fifteen people and turned Fort-de-France into a "smoking heap of ashes," with 90 percent of the city's inhabitants left homeless.[29] Just one year after the 1895 fire in Trinidad, in April 1896, fires ravaged Fort-de-France (again) as well as Colón, Panama. In August of that same year, when the crew of the HMS *Andes* went ashore at Port-au-Prince, Haiti, to help extinguish the huge fire in the center of that city, they were too late to save the Place Valliere ("one of the finest squares in . . . Port-au-Prince") and a number of adjacent buildings and warehouses.[30]

City (or town) fires were every bit as important for the urban places of smaller islands but at greatly reduced scales. Tiny clusters of buildings simply could not fuel the kind of fire that had devastated Kingston in 1882 and Port of Spain in 1895. Yet in relative terms, episodic Caribbean fires were perhaps even more damaging to small towns and cities than to large urban centers because they had the capacity to destroy one of these smaller places in its entirety. That is what seems to have occurred in Grenville, Grenada, in December 1886 when nearly all of the downtown area burned. About midnight on Monday the twentieth, "some evil disposed enemies of Mr. H. M. Douglas" set a fire in the upper part of his dry goods store on the main street running through Grenville. The blaze quickly enveloped Douglas's store and then was blown across the street, where it continued its destructive path and ultimately destroyed nearly all the commercial and government buildings in town.[31]

Smaller towns, especially if they were overshadowed politically and economically by a larger urban place on the same island, were often disadvantaged in having comparatively inferior fire-prevention infrastructures, increasing their vulnerability to arson as well as to accidental fires. In the case of Grenville in the 1880s, an earlier government decision to build a courthouse there instead of a modern water supply for the town was lamented after the 1886 fire. But Grenville lacked a good water system—despite a recommendation and survey that had been conducted in 1883—because Colonial Office officials in London simply would not allocate the necessary funds, since they already had done so for the capital of St. George's, across the island from Grenville. When a fire in St. George's in 1887 was extinguished quickly by "two strong jets" of water directed at the blaze, observers noted the ease with which that fire had been put out, in comparison with the blaze at water-deprived Grenville the preceding December.[32]

Intraisland rivalries for piped water were based not only on an island's environmental characteristics but also on the threat and fear of fire. A plan in the mid-1880s to provide Sandy Point, on the northern end of St. Kitts, with water to be brought down in pipes from the volcanic highlands met with vociferous opposition from influential (and fire-fearing) residents of Basseterre, in the south. The water source for both places was the Little River, in the mountains above Old Road in the center of the island. But that source was not dependable in winter, and owing in part to St. Kitts's northerly latitude, it ran dry in some years. Town fires in Basseterre had been extinguished—or at least combated—since 1874 with piped water from the central highlands, but the supply was always considered precarious. Water dependability, even if it meant no piped water for Sandy Point, was foremost in the minds of residents of Basseterre in the 1880s who recalled the enormous fire there in 1867 that destroyed £250,000 in property: "Since then we have taken more than a partial interest in our Water Works and supply, feeling assured that the protection of this town from fire . . . depends on a full supply of the element at the time needed."[33]

Modern water supplies elsewhere in the dry northern portion of the Antillean chain were obvious antidotes to that region's recurring aridity and its deleterious effects on people, crops, and animals. But water supplies were considered most important because of the threat of fires. A discussion of a new waterworks system for Nevis late in 1884 emphasized not how crops could be irrigated, sanitation improved, or thirst relieved but, rather, how the lack of reliable water systems to extinguish fires in nearby islands in recent decades had led to people being "driven from their homes by fire [and] exposed to the greatest privations . . . and destitute of common necessaries of food and clothing." Mass flight from town fires, the discussion continued, had occurred not only during 1867 in Basseterre, St. Kitts, but also at Pointe-à-Pitre, Guadeloupe, in 1868 and in Gustavia, St. Barthélemy, in 1852.[34]

Whereas city and town conflagrations drove people from their homes, burning buildings—even though they were obvious sources of danger and destruction—always attracted crowds of people. No two settings or scenes of Caribbean city fires in the 1880s and 1890s were the same, but as in other times and other places, onlookers invariably assembled to watch and often to help.[35] Often the fires occurred at night, their progress from small blazes to major fires indirectly aided by the inattention of a sleeping populace. When the bells, from churches or fire stations, or whistles sounded to signify a fire alarm, awakened adults and children poured out of their dwelling places to chase the flames. Caribbean newspaper descriptions of the time nearly always pointed out that the crowds at city fires represented "all classes" of the populace of a given island, a grouping apparently notable and obviously different from those at planned public events, where the proprieties of social segregation were more scrupulously followed. These spontaneously assembled groups of onlookers at nocturnal city fires usually had been roused from sound sleep, their reawakened mental states made all the more strange and unfamiliar by the presence of horse-drawn firefighting equipment, many shouts, moving silhouettes, and flames. Few onlookers had dressed themselves with appropriate convention beyond hastily putting on trousers and, if they owned them, shoes. The superficially undifferentiated crowds watching Caribbean city fires after dark in nightshirts and similarly informal attire were thus far different from the groups, composed of the same individuals, who might attend, say, a parade or public pronouncement in the light of day.

The excitement of following fires always swelled crowds in both large cities and small towns. A fire at 1:30 A.M. on Frederick Street in Port of Spain on May 19, 1891, which eventually burned down three buildings owned by the widow Hodge, inspired a typical clamor that awoke all neighborhood residents: "The cry of 'de feu! de feu! fire! fire!' was heard shouted by excited persons [followed by] the sound of many feet hurriedly rushing along the streets."[36] Five years earlier, on January 21, 1886, in Roseau, Dominica, the cries of "Fire! Fire!" similarly interrupted a moonlit Saturday night, and at first the flames appeared to be in the vicinity of the town's Anglican church. But the blaze was at the storage building on Wall House Estate about a mile from town. That distance did not deter onlookers on both foot and horseback from rushing to the scene in varying states of undress to watch the destruction of "the entire building . . . in one fearful blaze, as there were no means of subduing the flames"[37]

Despite the common experiences that briefly leveled the social differences among fire watchers, uniformed fire brigade members and police eventually took charge, and in the written reports that followed large city fires, the antics and bother that groups of onlookers created often inspired derisive comment. The destruction created by the huge 1895 fire in Port of Spain was said to have been exacerbated in the late afternoon when residents of the affected part of the city tried "to remove their possessions through a street choked with an excited mob." Later in the evening a detachment of U.S. marines performed in a reportedly exemplary manner when they "cleared the streets of a useless crowd which had done nothing but interfere with the efforts of those who were willing and anxious to do work."[38] Images of these events readily come to mind; although race and ethnicity were only rarely specified in these reports, terms such as "useless" and "excitable" were invariably reserved by authorities to describe members of the black working classes of the region.

In a few cases assembled crowds were said almost to cheer on fires, especially if the blazes were destroying the property of unpopular store owners, and reports of looting were common. In May 1897 a fire on adjacent sugarcane estates just outside Basseterre, St. Kitts, destroyed several acres of young cane sprouts. "Volunteers" were instrumental in eventually vanquishing the flames, but their efforts were thwarted by a number of onlookers whose behavior was described as "disorderly and almost aggressive."[39] The preceding year in Nevis, in the aftermath of the February disorders across the channel in St. Kitts, "a band of howling dervishes with all sorts of musical instruments" made road passage difficult to rural fires in the countryside. "They, of course, made no attempt to assist in extinguishing the flames, and if they did not fan the fire they certainly did much to sow the seeds of discord."[40]

There was not, however, a single type of group response or unvarying attitude among all of the crowds at Caribbean fires. Whereas it is understandable, given the region's social inequalities of the late 1800s, that hostile black crowds might encourage and support the destruction—by fire or any other means—of whites' property, it is more interesting and far more important that, in the great majority of reports about fires, the assembled crowds were said to have helped extinguish city fires, whether they were burning residential, commercial, or municipal property. Time and again, newspaper and government reports about "our good people" or "the always helpful onlookers" accompanied the written descriptions of city blazes. These texts can of course be interpreted as self-serving condescension by white authorities, writings composed by those anxious not to antagonize the black masses in times of especially combustible social atmospheres, and certainly that thought must have played a part in the oft-reported good behavior and helpfulness of fire crowds. But the written examples are simply too numerous and too specific to lead to any conclusion other than that the people of the region were genuinely willing, indeed eager, voluntary firefighters in the majority of cases.

Examples abound that tell of the helpfulness, hard work, and personal risks taken by onlookers in extinguishing these fires. At times the ad hoc efforts of bystanders put the official fire control personnel to shame. On January 21, 1887, a house fire in Basseterre "lighted up the whole atmosphere" and was visible for miles, but no policeman or fire brigade member could be found. The blaze soon was extinguished completely "by voluntary hands" some time before a sergeant and five men from the police station finally came strolling at a leisurely pace to the smoking remains.[41] Similarly, the action taken by neighbors at a chattel house fire in the Bank Hall area of Bridgetown, Barbados, on November 14, 1892, was a typical incident that seemed to occur almost weekly on the island. When the fire began, neighbors quickly "pulled down" several houses nearby so that the flames would not spread to the rest of the neighborhood, and they had the fire under control by the time the fire brigade arrived.[42]

When neighbors and onlookers pitched in to help save friends' houses, especially when the fires might spread to their own residences, their actions could obviously be categorized as exercises in self-interest. Yet in the great majority of occasions throughout the region, when rum shops, dry goods stores, and other commercial establishments were burning or threatened by fire, the public was there to assist. The example of the major fire in Castries, St. Lucia, in August 1899 illustrates the point. The fire broke out on the harbor town's wooden pier about 9:30 P.M. on Tuesday, August 8, and lasted until dawn. Although it destroyed the port's entire north wharf and the Barnard Sons store, the fire would have been even more damaging had it not been for the joint efforts of the local fire brigade, volunteers from the military installation near Castries, crew members of the visiting British steamer *Chiswick*—in port to take on coal—and, notably, dozens of St. Lucia's women coal carriers: "The activity of the women was particularly noteworthy." The *Chiswick* crew members and coal carrier women together formed bucket brigades, hauling water throughout the night to keep the coal piles wet.[43]

A similar show of "bravery, pluck and determination in both sexes, young and old," in fighting a government warehouse fire in Portsmouth, on the northwestern coast of Dominica, exemplified a communal and apparently classless spirit in Caribbean fire fighting in October 1899. The warehouse contained puncheons of rum along with quantities of mining fuse, tar, pitch pine, and other combustible material that, despite the firefighting efforts, altogether produced "a spectacular blaze." The good faith displayed by residents of Portsmouth might be categorized as custom that had somewhat recently become law. Unlike some of the larger islands, Dominica had no organized fire brigade until a voluntary unit was organized in Roseau in 1898. Until that time, fire fighting on the island had been vaguely considered a police responsibility, although the fire menace by late 1893 had become so acute that a Dominica court decreed "that putting out fire . . . was the duty of every citizen."[44]

Kerosene

Two weeks after the fire of 1895, the Port of Spain city coroner ended the formal postfire inquest by concluding that although the place of origin of the conflagration seemed certain, the cause was unknown. He simply could not be sure whether it was "due to accident or to the hand of some incendiary."[45] Such inquests always followed major, and some relatively minor,

Caribbean city fires, with hearings that included judicial officers, the police, fire brigade officials, property owners, and the local agents of the relevant fire insurance companies. Witnesses—usually police constables who had first seen the blazes, owners of the damaged property, and passersby who had observed significant or suspicious activity prior to the fires—provided oral testimony. Whereas newspaper accounts of these inquests often portrayed the social and environmental atmospheres surrounding these events, whether they be, for example, the recent dry weather or an ambient "spirit of incendiarism," the hearings themselves nearly always focused on the particular fire's original ignition.

"Reasons" for the igniting of the innumerable fires in the wooden shantytowns of eastern Caribbean towns and cities in the late 1800s were hardly mysterious. The newspapers of the era fairly bristled with reports describing the innumerable house fires started either by children playing with matches or from homemade kerosene lamps exploding or being tipped over. The latter—sometimes referred to as "dunkey pumps"—were fashioned from discarded tins partially filled with kerosene, their covers precariously perched on top of the containers, through which wicks were passed. The interior of every small house, as in Barbados in 1886, was lit with one of these dangerous devices that posed constant fire hazards to their occupants: "Owing to the slightest accident the lamp is overturned and the inflammable oil ignites."[46] The kerosene accident that killed a young woman named Floretta Alleyne in Bridgetown in November 1881 was an event often repeated in Barbados and nearby islands; while Alleyne was filling a tiny kerosene lamp, the fluid ignited and set her dress afire. Although she received professional medical attention, probably more rapidly than most, she died from her burns the following morning.[47]

Kerosene first came to the region no later than the 1860s, after a young Canadian, Abraham Gesner, developed the technique of extracting kerosene (which he also named) from asphalt, applied for a U.S. patent (1854), and thereby began an international industry for the substance whose principal use was that of an "illuminating oil."[48] Thereafter kerosene oil came to the Caribbean, sometimes from England but mainly from the United States, in shipboard tanks. By the late 1870s the imported oil, used as an illuminant in both domestic and municipal contexts, had revolutionized lighting in the islands and was an important import staple. In 1882, Dominica, for example, imported 28,323 gallons of kerosene, roughly the average aggregate quantity Dominicans imported throughout the remainder of the decade.[49] The larger and more populous islands by this time imported far more, and local colonial officials quickly realized opportunities to capitalize on the widespread and increasing demand for the new lighting oil by imposing import duties on it. By the late 1880s the government of Trinidad realized £16,000 per year in kerosene duties, a tax "that weighs mainly on the masses."[50]

Kerosene provided more reliable light in the islands than ever before after the sun set at 6:00 each evening, a remarkable improvement when compared with what kerosene replaced: tallow candles, wooden torches, or in many small houses, rags dipped in animal fat that burned and sputtered, emitting only pale, flickering light. Indeed, prior to the common use of kerosene lamps, an inky blackness was said to envelop the mountainous and rural zones of the islands at night; in village areas on nights without moonlight, the pervasiveness of the darkness inspired comment from more than one visitor. Darkness prevailed in the urban areas, too. Before the introduction of street lighting to Barbados late in the nineteenth century, "Bridgetown at night time was . . . a city of darkness in a literal sense."[51] Similar conditions were even more obvious in the region's smaller urban locales; in the early hours of March 4, 1881, the Royal Mail steamer *Tiber* sounded its horn in the vicinity of the terminus of the Roseau, Dominica, pier. But there was no torch-bearing police constable to meet and direct the vessel as usual, so the ship had to wait until dawn to dock.[52]

When street lighting with kerosene and "town gas" (an illuminant distilled from coal) was first introduced to the eastern Caribbean, to Bridgetown in January 1876 and to Port of Spain two years later, it represented one more element reinforcing urban-rural differences and interisland contrasts, and it was hailed as, among other things, a promising deterrent to petty crime. The introduction of street lamps did not occur everywhere at once, coming a few years later to the smaller islands. The first illumination of the market square in St. George's, Grenada, in April 1882, for example, was achieved by kerosene lamps attached to the tops of four cast-iron lampposts projecting light for distances of thirty paces from each post. The "very satisfactory" effects of these new lights inspired the Grenada government to import seventeen more sets of posts and lamps later in the year for other parts of the town.[53] But enthusiasm for kerosene town lighting was not unanimous in the region, and by the time electricity began to replace it in Port of Spain early in the 1890s, the oil lamps were lit only on nights when the moon was not out, and they were decried as providing "scarcely . . . more light than so many candles."[54]

The brief era of lighting streets exclusively with kerosene in the towns and cities of the eastern Caribbean (roughly the last quarter of the nineteenth century) always involved an accompanying and continuous fear of accidental fire. At dusk in the 1880s, lamplighters—their duties sometimes performed by police constables—from St. Kitts to Tobago usually ventured out to light the wicks of the several kerosene lamps atop the metal stands around the towns. Far from inspiring romantic reflection, these men were often condemned for their unreliability and carelessness that led to lamp explosions and worse. In September 1882 the "Public Kerosine Lamp" across the street from the local newspaper office in Roseau, Dominica, exploded, setting the surface of the street ablaze and endangering nearby buildings. In May of the following year in Roseau, police were summoned to extinguish a similar fire, leading to calls for the dismissal of Mr. Lafraicher, the town lamplighter.[55] By the mid-1890s kerosene street lamps were considered, as in Antigua, constant sources of "danger and discomfort." Kerosene street lamps were eventually abandoned there, especially with the news of the advent and promise of the electric light.[56]

Most West Indians, however, soon realized that the fire dangers accompanying the new illuminating oil had far more to do with the varying quality of different batches of imported kerosene than with the personal attributes of those charged with lighting street lamps each night. In the early days of kerosene refining everywhere, not only for oil sent to the Caribbean but for that used in North America and Europe as well, if the kerosene contained too much gasoline or naphtha, "the purchaser's attempt to light it could be his last act on this earth."[57] As kerosene came into increased use in the eastern Caribbean in the 1870s and 1880s, the dangerous differences in quality of imported kerosene often led to accidents and explosions, less from street lighting than from among the small lamps that lit individual dwellings. An observer in Dominica in mid-1882, typical of many, asserted that there were two kinds of kerosene in use in the island; the one that was more flammable—and therefore cheaper—"enters largely into our use, and keeps us always in fear."[58] Later that year the local newspaper editorialized about the dangerous grades of kerosene commonly used in Dominican dwellings among people of "all classes" and urged the importation of higher-quality "Home-use Oil," said to be popular in Barbados.[59]

Widespread claims about fraudulent importers or grasping shopkeepers adulterating kerosene with pitch oil, combined with an increase in kerosene-related house fires, led concerned officials to take measures to regulate the quality of the imported oil. A bill in Antigua in March 1883 was intended to regulate the importation of kerosene there. By March 1886 Barbados had adopted legislation "to impose a high test on the kerosine imported for sale."[60] But a lack of standardization from one island to the next could, and apparently did, lead to transshipments of lighting oil considered dangerous in one place to islands where rules were more lax. Early in 1886 a shipment of oil said to be "very combustible" was rejected by importation authorities in British Guiana and then shipped to Trinidad, where "it has no doubt duly arrived and gone . . . into consumption."[61] So grave concern prevailed among those who desired more light at night and who therefore bought kerosene—which, by the 1880s, included nearly everyone living in the islands—as to whether the substance would explode and seriously injure them or their family members. By 1885 (presumably effective) newspaper advertisements appeared in Trinidad telling of the virtues of Bush & Denslow's Premium Safety Oil, "the only Oil in the United States that has ever been officially endorsed by the New York Board of Fire Underwriters."[62]

Related newspaper ads proclaimed the safety of imported, factory-produced glass and metal kerosene lamps similar to those comprising so much of the array of "kerosene goods" that were staples of shops and stores in the United States of the era.[63] But unlike in North America, the incomes of members of the black working class in the Caribbean were so pitifully low that most were unable to afford even the cheapest manufactured glass objects.[64] The price of kerosene itself—one shilling per gallon in St. Lucia at the turn of the century—was so high that common people could afford to purchase it in only tiny quantities, usually refilling their homemade lamps every few days at local shops for a penny or so.[65] Consequently, the combination of small, unsafe, homemade "lamps" for kerosene that were often fueled with low grades of oil, the continuous carrying back and forth of small quantities of extremely flammable material, and the combustible nature of the region's small wooden dwellings together created a frightfully dangerous fire hazard common to all of the towns and cities of the region. In commenting on possible reasons for the fire that destroyed much of Fort-de-France, Martinique, in June 1890, the *Voice of St. Lucia* asserted that the poor people of the region "use petroleum more or less recklessly," which of course provided a pat explanation for the Martinique tragedy as well as for the prevalence of urban fires all through the islands.[66] But government officials could be every bit as clumsy with kerosene, such as the policeman who overturned a lamp at the new Port of Spain police headquarters in November 1881, thereby starting a fire that destroyed the entire building.[67]

There is no evidence that the Trinidad oil industry, in its embryonic stage during this period, ever provided kerosene for the local market at the turn of the century. The definitive written history of this enterprise points out that asphalt gathered from the pitch lake at La Brea was important in road covering, roofing, cementing, and the like, but that the conventional uses for petroleum from Trinidad in producing, among other products, "illuminants" did not occur until the 1920s or so.[68] Yet the advent of the kerosene age in the Lesser Antilles and elsewhere in the world late in the nineteenth century did directly affect one small but significant socioeconomic group in the eastern Caribbean. Beginning in the early 1800s, American whalers had been active in the waters of the Grenadines. Their presence led to the development of local whaling, an extractive enterprise manned exclusively by seamen from the Grenadines that produced close to 1,000 barrels of whale oil a year in the 1860s. When cheaper petroleum-based lighting oil came to the islands in the 1870s, the demand for, and therefore the production of, whale oil from the Grenadines and elsewhere declined precipitously.[69]

In February 1896 a display of lighting by electricity at the Bridgetown waterfront demonstrated the new marvel to members of the Barbados legislature. Yet it was not until 1910 that Bridgetown was illuminated by electricity.[70] Although street lighting with electricity came several years later to Barbados than it did to some of the nearby islands, these dates correspond roughly to the experimentation with and adoption of electric lighting everywhere in the region. By the early 1880s, when news of the new phenomenon was widespread, newspaper commentary throughout the islands stressed the seemingly safe (yet expensive) nature of electricity when compared with kerosene. In St. Lucia late in 1892, the Castries Town Board was considering both the electrical "Rochester light" and the "Dietz tubular lamp," as well as an improved kerosene lamp, for possible adoption for the town's streetlights.[71] Electricity lit some of the streets of Port of Spain by 1894, accompanied by an aesthetic shortcoming in the virtual forest of unsightly wooden poles on which the electricity cables were hung.[72] All streets in Port of Spain were illuminated by electricity by the turn of the century, although the municipal authorities were still negotiating in 1901 with several companies that provided electric lighting.[73]

Although official commentary about the use of kerosene among the working peoples of the eastern Caribbean in the late nineteenth century focused nearly exclusively on the innumerable house fires associated with it, the provision of reliable light at night by kerosene for common people surely represented a subtle yet pervasive and important element of social change. Poor people could converse, read or learn to read, and assemble in groups, even within the confines of small village areas, now that their activities were accompanied by light for a longer period than the twelve hours of sunlight in each day. More opportunity to discuss local events, church matters, self-help strategies in light of common socioeconomic circumstances, and life in general now presented themselves thanks to controlled kerosene fire. The origins of most secondary schools in the English-speaking Caribbean islands, a development that took place after about 1870, were not directly caused by the kerosene that arrived about the same time.[74] But many students subsequently must have read at night by the light from kerosene lamps. The organization and planning of events, conspiratorial and otherwise, by local peoples must have taken place by the same light. Yet these speculations must be tempered with the constant (and far better reported) threats of accidental fire that kerosene always carried with it. In rural villages of Antigua in the 1890s, "early in the evening the kerosene lamps would be put out for fear of fire."[75]

Electricity by no means eliminated kerosene lighting in the islands after the beginning of the twentieth century but, rather, pushed it more and more into rural and remote areas. Late in 1907 some of the "subscribers" to the electricity system in Roseau, Dominica, continued actively to employ kerosene lamps to augment the new lighting system.[76] Kerosene lighting continues early in the twenty-first century to play important roles in rural zones of the Lesser Antilles, in the very small islands, and as emergency sources of lighting in the autumn hurricane season everywhere in the region.

Accidents and Arson

Kerosene lamps did not burn all night. Similarly, charcoal cooking fires had to be reignited each morning. A small early morning blaze that endangered one of the stores in St, George's, Grenada, early in 1883 was traced to a prevalent (and, strictly speaking, illegal) practice of the time, that of "sending children from one place to another along the streets in the early mornings to fetch live coals or burning sticks for raising a quick oven or kitchen fire."[77] This practice of trading or "borrowing" fire between one small wooden house and another was common in Caribbean countrysides and the poor urban areas, as it avoided the squandering of too many wooden matches that, although becoming common by the end of the 1800s, were still, for impoverished people in the midst of economic depression, very expensive.

Despite the widespread use of matches in Europe and North America by the late 1800s, traditional fire lighting techniques still prevailed in some Caribbean countrysides. Until the 1890s in Antiguan villages, flint and metal—when struck together—produced sparks that lit tiny flames in cotton that had been wadded into small containers, devices crafted by village tinder box specialists.[78] Yet the imported wooden match, or "lucifer," first invented by British chemist John Walker in 1827 and subsequently developed into a mass-produced commodity in Austria, Germany, and Sweden, was making important inroads in Antigua and elsewhere in the islands. As production costs and prices fell for imported wooden matches, they came increasingly into the possession of even the poorest inhabitants of Caribbean cities and villages. Thereafter "there was no control by cost, skill, or tradition on who had access to fire," and although it would be difficult to quantify the assertion, the proliferation of wooden matches in the late 1800s, along with the growing use of kerosene, almost certainly increased the number and frequency of Caribbean fires.[79]

The opening of match factories in British Guiana in 1883 and in Trinidad and Barbados soon thereafter had the overall effect of putting more matches—and thereby the potential for more fires—into local hands. The Red Star Match Factory, built in Bridgetown in 1887, manufactured a product that was less susceptible to dampness than were the German imports, and Red Star boxes held 100 matches, whereas only 60 matches came in the imported ones. A match factory was also started in Port of Spain in the same year, and it produced not only cheaper and better matches than those imported but also a few local jobs. The assembly of small boxes, which when filled with matches cost customers one penny apiece, was accomplished by local Trinidadian women. Production was curtailed, however, in late October 1888 when most of the Trinidad match factory was destroyed by fire.[80] Useful as matches were for lighting cooking fires or burning bits of rubbish around small houses, they were obviously dangerous when controlled by children, in both urban and rural areas. In September 1889 near Toco, in the far northeastern corner of Trinidad, an old woman named Charlotte Hope was minding two of her grandchildren and gave the older one a crab to eat. The child then took a box of matches from under her grandmother's pillow ("a common hiding place for matches among these people") to light a crab-roasting fire. The child mishandled the lighted match and dropped it onto the floor, setting the house afire in seconds. Before any help could arrive, "the unfortunate woman was buried under the blazing roof and burnt to death."[81]

The more numerous reports of accidental fires in small dwellings in urban areas provided a fund of evidence about the construction materials that composed these houses and about the character of the dwellings themselves. The fire that destroyed a tiny house in Port of Spain in October 1898 reportedly left "the premises . . . burnt to ashes, only the galvanize being left."[82] Galvanized iron was the common roofing for the city's notoriously run-down barracks ranges, rows of partitioned rooms behind the stores and homes that fronted the main streets. These elongated wooden backyard sheds, each usually with a common back wall, were subdivided by flimsy partitions into as many as ten different rooms, each about ten feet square. Beds stood at the back of each room, washing was accomplished at communal standpipes, privacy was afforded only by paper partitions, and latrine facilities were a common pit "surmounted by a wooden hut generally divided into two compartments . . . with doors so warped and hinges so rusty as to be of little use."[83]

The galvanized iron roofing common in Port of Spain was less evident in the descriptions of small, makeshift city houses in the other islands, where occupants occasionally used small beams, flooring, and shingles fashioned from tropical hardwoods but more often built their dwellings from imported pitch pine. The principal building materials for the tiny working-class houses throughout the region always were augmented by scraps of metal or parts of old boxes or buildings found in the dump heaps around city areas and plantation yards. Whatever the building components were, the small dwellings nearly always burned furiously after they aged. Observers in Barbados on July 16, 1888, in only one example, noticed the night sky "brilliantly illuminated" and followed "the natural impulse which the sight of a fire always exercises" until they reached "the flames . . . proceeding from a wooden house . . . in that unsightly nest of small buildings all of wood and some of them in the rottenest state imaginable, aback of Chapman's Street."[84]

Official attitudes toward workers' houses led to town planning policies that actually increased rather than reduced the potential for urban fires. In beholding the unsightly and crowded dwelling areas for the poor, colonial officials thought they were seeing not only painful eyesores but also the physical bases for the disgusting, immoral, and even criminal behavior among these neighborhoods' inhabitants. The unsightly dwellings had their social equivalents, in official eyes, in the noisy quarreling and obscene language emanating from the crowded barracks in city yards and alleyways. So one had to look no further than the built environment itself for explanations for the objectionable behavior of bands of "roughs" who stole fruit from hawkers' trays, grabbed hats on streetcars, disrupted public celebrations, and worse. Official attempts to confine, contain, and insulate respectable citizens from the pockets of urban dwelling areas for the poor led to zones of unwanted (by officials) city neighborhoods that were out of sight from city centers and closely packed with small, potentially combustible houses. These policies, practiced on every island, seem to have created ideal conditions for the spread of urban fires on dry and windy days. Early in 1894 the Castries Town Board, for example, took a stand to confine the construction of wooden "shanties or unsightly houses" to "certain parts of the town," thereby reducing the "incongruity between mean houses and fine streets."[85] A similar confinement policy in Bridgetown, Barbados, four years earlier helped to explain why no fewer than 266 tiny houses were packed into three acres at Church Village; these wooden dwellings were separated by alleyways, the widest of which was only thirteen feet.[86] Similar policies throughout the region led to numerous outbreaks of the kind of fire that raced through several small wooden houses on Tanner Street in St. John's, Antigua, in mid-August 1890 after a cooking fire blew out of control: "And the wonder is that fires from this cause are not of more frequent occurrence."[87]

White officials also surmised (although actual proof was very rare) that they needed to look no further than these ugly clumps of wooden shacks, tucked away in confined and unwanted corners of British West Indian cities, to locate those responsible for the ongoing menace of urban arson. Few prospects were more actively dreaded by officials and property owners than a wave of maliciously set fires that, under the wrong weather conditions, could destroy much of a city. From 1880 to 1886, of the 123 fires reported in Port of Spain, no fewer than 53 were considered arson.[88] If flames from either arson or accidents could destroy all of downtown Port of Spain, as in March 1895, one could easily imagine what they could do to smaller places. Reports of the attempt to set fire—with the use of kerosene and coconut husks—to a mercantile establishment on the main street of Scarborough, Tobago, in April 1888 led to the professed concern that someone was apparently trying to burn down the entire town.[89]

The local newspapers heightened fire-related tensions with their own inflammatory rhetoric. "Fire fiends" and "fire bugs" and "incendiaries" were responsible for city blazes from the Leewards to British Guiana. When inexplicable rural fires occurred, "Captain Moonlight" roamed the countryside. Usually such terms were applied only to local incidents, but on occasion they extended across the region. When the news of a series of sugarcane fires and the burning of a slaughterhouse in Barbados in 1891 was combined with notices about city fires in Trinidad and the burning of the match factory in British Guiana, the *Barbados Globe and Colonial Advocate* pronounced the situation "ominous" and worse: "The Fire King seems to have ascended his throne in the West Indies this year."[90] Even fires in distant places were, on occasion, considered arson; the newspaper in Dominica in June 1894 suggested abruptly that a recent series of fires in Panama "can hardly be on every occasion the result of accident."[91]

The availability of kerosene provided incendiaries the wherewithal to destroy fields and buildings as never before. Investigators of suspicious or mysterious city fires sought, and often found, evidence such as spent matches along with sticks or rags soaked in kerosene. The latter could be stuffed into the latticework in wooden doorways, underneath buildings, or through cracks in windows and then lit with a match, so arsonists needed no entry to set a building ablaze. In August 1899 a Bridgetown police detachment discovered attempted arson at a store in High Street. A jar containing a lighted candle had been placed inside a bucket half full of kerosene in a room where five five-gallon cans of the lighting fluid had been opened and poured onto the floor. The police extinguished the candle, thereby preventing what might have been a major fire. They also apprehended the arsonist, "old convict" Robert Bradshaw, by identifying stolen property from the store. Such an arrest was indeed a rarity because arsonists were almost never caught.[92]

Although the means to set a fire were reasonably straightforward, motivations were varied and complex. Reasons for incendiarism varied from pique for trivial incidents to overtly criminal behavior. No doubt the infectious character of maliciously set fires, the oft-described "spirit of incendiarism," helped create atmospheres conducive to arson. Yet except for rare lightning strikes, which occurred sometimes in the highlands of the volcanic Windwards, people, not "atmospheres," ignited Caribbean fires. Feelings of outrage on the part of members of the black working classes facilitated arson, and these general feelings were well understood (but often unspoken so as not to provoke more fires) by members of the elite who might suffer the consequences. In any case, precautions were sometimes taken against providing the wherewithal for destructive fires. Warnings against the importation and sale of dynamite "in the Portuguese shops as if it were saltfish or tallow candles" in St. Kitts early in 1885 noted that the material could be used "as an engine of vengeance."[93] But exploding dynamite and setting fires as acts of outrage were not random acts, and vengeance was almost always directed against specific targets. As only one example, in a "dastardly outrage" on the Mount Craven Estate in Grenada in November 1893, a "ruffian or ruffians" cut the throats of and disemboweled four of the estate's mules "and then set fire to the megass house."[94]

Arson inflamed and thereby publicized a variety of other conflicts and frustrations in the depression years. A fire set inside the Fairchild Street railway office in Bridgetown, Barbados, in October 1891 was "supposed to have been the work of an incendiary who in search of money was disappointed, and adopted this method of shewing it."[95] In 1896 in Port of Spain a series of intentionally set blazes around the city might have been the work of unemployed men desperate for wages. They had been "reduced to the verge of starvation" and possibly were hoping for more of the "phoenix effect" that the great fire the year before had given to carpenters, masons, and others in the city by providing rebuilding jobs.[96] Fires in some of the small huts near the army barracks above Castries, St. Lucia, in April 1898, as "everybody knows . . . were not accidental, but arose out of quarrels between soldiers and the camp-followers infesting the neighbourhood."[97]

But arson was not always an outgrowth of the bitterness that accompanied the regional inequalities in color and class. Postfire inquests in Caribbean towns and cities focused, as often as not, on seeking evidence to confirm whether fires were set deliberately by property owners to collect insurance payments. In September 1896 a Barbados newspaper expressed "no surprise" over the cabled reports of a series of fires in Port of Spain houses covered by insurance. Perhaps the Barbadians were overly suspicious of insurance-related arson because of its prevalence in the white suburbs southeast of Bridgetown. In 1893 a fire there attributed to an elderly white couple (subsequently acquitted because of the absence of ironclad proof) was almost expected, since "a fire at Hastings, sometime in the Christmas holiday, seems to be as certain as Christmas itself." Five years later, in early 1898, the rumored Barbadian propensity for insurance-related arson had taken an ugly turn. Police investigations of a string of suspicious fires exposed "a new and alarming development." It seemed that owners of failing businesses and unrented houses had procured the services of "professional fire fiends" who were said to torch buildings on demand for those same owners "with the object of pocketing the money of the Insurance Company."[98]

Despite the recurring, publicized offers of monetary rewards for "fire fiends" and "incendiaries" throughout the region in the late 1800s, those responsible for the innumerable acts of arson were only rarely identified or brought to trial. The luminous, overt, public character of arsonists' blazes found their counterpoints in the darkness and obscurity of closed-mouth collusion that allowed the perpetrators nearly always to go undetected. The brief comment following a deliberately set fire at Langford's Estate in Antigua in June 1884 echoed similar refrains on all the islands when cases of arson arose: "That such an act of Incendiarism can be committed in such a small community without detection, shows how much combination existed in perpetrating it."[99] But the term "community" in Antigua and the other islands really described only the proximate physical circumstances in which people lived. The asymmetries of and barriers between different peoples help to explain far better why so few fire starters ever were apprehended.

Yet it would be wrong to imply that all fin-de-siècle British Caribbean societies were starkly bipolar. Ethnic differences and subtleties from one place to another occasionally had their expressions in terms of, among other things, arson accusations and arrests. The all-night fire on the Castries wharf in August 1899 that brought together St. Lucia women and British sailors as common members of bucket brigades was generally regarded as having been set by immigrant Barbadian "wharf scum," men who recently had come to St. Lucia from Bridgetown to look for work and who were regarded by the local officials as lazy, sinister, and criminal. Similarly, the few reported arrests of incendiaries in the region at the time seem always to have occurred in settings of ethnic diversity, where a member or members of one group turned in someone from another. In Dominica in September 1886 an indentured Indian, identified by black neighbors, was arrested for burning down two buildings on Soufrière Estate near Roseau.[100] In December 1897, to cite one of many Trinidadian examples, four young black men were arrested for setting a Chinese shop ablaze in the Caroni district in the west central part of the island.[101] Had the Indian set his fire on one of the coastal estates of British Guiana or the blacks committed arson in one of the rural parishes of Barbados in those same years, it seems nearly certain that they never would have been apprehended.

The incidence of fires in urban—as opposed to rural—areas in Caribbean history is probably overreported in conventional archival sources for the simple reason that the colonial officials and newspaper reporters, on whose written records we depend, were more directly affected by city fires than by blazes in the country. But in some cases the "bush fires" of the countrysides threatened to invade towns and cities, especially in times of drought. "Rural" and "urban," however, are not always useful constructs because they can blur the reality of the continuous human movement and interaction from one area to another, especially in small places. These connections were illuminated in early 1899 in Barbados when a sugarcane estate fire helped to highlight urban impoverishment. On February 14 the city fire brigade came quickly to combat an early evening blaze at Bay Plantation, just outside southern Bridgetown, but poor people from the city already had preceded the firemen, taking advantage of the confusion to satisfy their hunger: "Men, women, and boys were seen on the highways leading from the field eating and bearing away the canes."[102]

4

Forestry and Bush Fires

DESPITE THEIR TINY LAND AREAS, THE ISLANDS of the Lesser Antilles were not comprehensively monitored and thoroughly traveled by the town-dwelling colonial authorities in the 1890s. The well-publicized hikes and outings by British governors or administrators to interior mountains or to the small islets that were dependencies of larger islands were uncommon events and rarely attempted during periods of high seas or impassable roads, conditions often prevailing in the latter half of each calendar year. Therefore the archival records and reports—the academic evidence upon which most of the region's written history has been based—were mainly reflections by those whose daily activities were confined to towns and cities. Rural areas, in contrast, usually went relatively unappreciated, infrequently encountered, and poorly understood.

To be sure, officials easily could and did visit the sugarcane estates on the low-lying islands, which had high degrees of internal accessibility; Barbados provided perhaps the most obvious example. Annual production reports from these same plantations formed the basis for much of the official recorded information about the region at the time. But on the more mountainous islands, town-based commentators often marveled, for example, at the groups of country market women who seemed to appear out of thin air each Saturday morning, their baskets brimming with fresh fruit, eggs, and harvested root crops. The rural garden plots where this produce had been cultivated, even if they were in the next valley across the mountains, were all but unknown by colonial officials. This ignorance, in turn, no doubt buttressed the widespread myth of West Indian tropical fecundity that had foodstuffs hanging from the boughs of every tree, the image so eloquently (and inaccurately) perpetuated by historian James Anthony Froude during his trip to the region in the late 1880s.

The town-based ignorance about country districts in a place like Dominica was certainly understandable. Except for a few cultivated valley areas, Dominica offered a seemingly impenetrable mountain landscape of forested peaks and forbidding ridgelines barely touched by the trails extending inland from coastal roads; these interior "traces" were notoriously overgrown and often impassable from rain-induced mudslides and landslips. So it is not difficult to understand that in the rainy season in September 1880, the newspaper office in Roseau lamented the "great difficulty . . . in obtaining intelligence from the country districts, of the state of the weather and the prospects of the forth-coming Crop." But ignorance of the countryside was regionally widespread. Even in nearby and reasonably flat Antigua, the lack of day-to-day knowledge about the island's interior meant that residents of St. John's simply did not know what to expect when problems arose in the island's rural parishes. Early in 1895 the twin evils of severe economic depression and a pitiless drought on Antigua had led to "unprecedented severity" in local poverty that was endemic "in some of the country districts of which we residing in the city know literally nothing."[1]

Rural fires, in Dominica, Antigua, and the other islands of the region, were among the few events in which country districts commanded the attention of record-keeping town dwellers. Everyone knew, if for no other reason than because of the seasonally smoky skies or from occasional glimpses of faraway flames at night, that rural burning occurred from time to time in country areas. But these fires had to be of sufficient severity or proximity to the towns to warrant attention. During the dry season in early 1899, for example, the "bush fires" on Dominica, which were set annually by local cultivators to clear land, threatened nearby village areas, and flames were "seen night after night raging at intervals from the south to the north end of the island." In the same year similarly arid conditions in Trinidad helped to explain the widespread bush fires throughout the heavily settled western part of the island, including some that appeared in the hills behind Port of Spain and threatened outlying city buildings.[2]

Urban-based inattention to most rural fires in the eastern Caribbean at the time perhaps diminished the general awareness of how fire was shaping local insular landscapes, processes intensified because of the sugar depression. In islands that already were marginal sugarcane producers, abandoned cane fields and adjacent areas informally controlled by the descendants of former estate workers were being transformed into grazing areas through the use of fire. Antigua and Dominica again provided examples. On the former island, peasants burned lands that had originally supported small trees and thus produced a drier overall look to an already drought-prone island by reducing the possibility of woody growth. Black rural dwellers set pasture fires every year on Antigua in the late 1800s to destroy ticks and other insect pests as well as to remove the "old coarse dry grass in order to produce a fresh growth upon which cattle will feed." These burnings had important side effects, as they destroyed "young seedling trees by hundreds" and were able to "undo all that nature is striving to do towards the clothing of our hillsides with desirable timber."[3]

Fire's overall geographical effects were even more important in rural Dominica. Land clearance by small-scale subsistence planters and herders there in the 1890s usually involved the setting of subsequently uncontrolled dry-season fires that damaged forest areas as well as the trees on neighboring cacao and lime plantations. In some of the leeward areas of Dominica, in the same places where coffee had been cultivated in the 1870s and later abandoned, repeated burning year after year left the soil "burnt and bare," and the landscapes now consisted only of "barren wastes of rocks" rather than the expected secondary forest growth. This seasonal firing of field and forest in Dominica had then apparently set in motion important geomorphic processes. Rootlets and other vegetable matter in the soil were reduced to the point that little was left to protect its susceptibility to sheet erosion when the heavy June-to-October rains fell, and the formerly fertile soil of the island was now "washed to the valley or sea." Dominica was not the only place in the northeastern Caribbean where this destructive, fire-induced erosion process was occurring, as it was also reported for both Montserrat and Anguilla.[4]

Although British West Indian colonial authorities of the late 1800s noticed and knew relatively little about the fires set outside town and city boundaries, their ignorance did not prevent them from disliking countryside bush fires in their many forms. This dislike, in turn, had far less to do with the fires themselves than with those who set them. Town dwellers and plantation workers, who were within easy reach of colonial authorities, came readily under direct official control. Geographical inaccessibility, on the other hand, bred contempt. Villages far removed from colonial capital towns and the few isolated settlements in the semiwild highland areas were inhabited by people who could not so easily be "registered, counted, taxed or watched."[5] Their sense of freedom was readily expressed by, among many other things, their extensive and freewheeling use of fire, which represented visible, recurring, and perhaps momentarily exciting celebrations of their relative independence. Although the late nineteenth-century Caribbean colonies had belonged to Britain for more than two centuries, some parts of the islands still required administrative vigilance lest their inhabitants relapse into the barbaric and savage behavior better known in the newer parts of the empire. One way to prevent such a slide was to monitor and curtail cultural practices that involved uncontrolled burning. It was well known among colonial authorities in the West Indies, as in other parts of the world at the time, that "sovereignty over land demanded sovereignty over fire."[6]

Imperial Forestry in the Caribbean

In April 1882 Daniel Morris, the director of public gardens and plantations in Jamaica, dispatched a lengthy, handwritten report to the head of the Royal Botanic Gardens in London. Morris's report expressed alarm about the reckless forest destruction in Jamaica in general, condemned the wanton use of fire there in particular, and set in motion plans for what would eventually become the only comprehensive forestry survey of the British Caribbean the region had ever known.[7]

Morris based his report on recent local records and hearsay as well as his own field observations. He pointed out that widespread land clearing in highland interior Jamaica for small-scale agriculture, a system that fell vaguely under the authority of the colony's Crown surveyor, saw roughly 30,000 acres of native forest cut and burned every year. Almost all of the cleared timber was burned on the spot, very little was salvaged for commercial use, and forest replanting was nearly nonexistent. All of this destruction yielded nothing more than a few cleared plots of "yam and other ground provisions." Morris himself had witnessed one instance where probably £100 worth of "beautiful mahogany" and other hardwoods had gone up in smoke "to plant an acre of yams." In another case, he knew of a fire that had burned out of control and destroyed fifty acres of forest "merely to supply a negro cultivator with half an acre . . . to plant cocos and yams." Morris condemned this "enormous waste" as not only economically unsound but also detrimental to Jamaica's environmental well-being because both drought and massive soil erosion resulted from this "reckless and lawless" denudation of the island's forest slopes. Morris's concluding recommendation to save Jamaica's mountainous interior from further ruin was a condemnation of wasteful burning. Specifically, the colonial government, in Morris's view, should take legal actions toward "the prevention of forest fires. These fires are very frequent in remote backlands, and are chiefly caused by the careless use of fire on negro provision grounds."[8]

At Kew the report was endorsed strongly and acted upon by Sir Joseph Hooker, the director of the Royal Botanic Gardens, who in August sent a circular letter to all of the governors of the British Caribbean colonies. In his memorandum Hooker reiterated and expanded on many of Morris's observations and wondered if the poor state of conservation was a regionwide—not simply a Jamaican—shortcoming. To answer the question, Hooker recommended to the various governors a general fact-finding visit to the region and "the employment of a trained Forest Officer to examine and report on the Forest Question in all the British West Indian Colonies." Hooker was not in a position to finance, only to recommend, such a sojourn, and he left it up to individual colonies in the region "to contribute towards defraying the expense of obtaining the services" of such an individual.

In his letter to the West Indian governors, Hooker condemned, as had Morris, the wasteful environmental practices in interior Jamaica (and, by implication, elsewhere in the Caribbean) and cited possible solutions that the governors probably found attractive. He pointed out that any country's forests, every bit as much as the land, were part of its "natural capital" and that it would therefore be "impolitic" to let those forests go to waste. But the situation in Jamaica was not simply wrong; it was "incredible": "The system of 'provision ground letting' so obviously stands self-condemned that it is almost unnecessary to discuss it. To allow the negroes to destroy acre after acre of woodland . . . seems economically ridiculous." To allow these events to recur, Hooker continued, amounted to "scandal." A partial solution seemed obvious: "On moral grounds alone the migratory and squatting habits of the negroes should be as far as possible arrested." Hooker concluded that the West Indian situation, with the aid of a properly trained scientist, might be turned around if local governments were apprised of some of the recent innovations in modern forest management, and he cited an article on the subject recently published by Colonel G. F. Pearson at the French Forest School at Nancy.[9]

Forestry author Pearson originally had made his mark (by dealing successfully with tropical forest fires) in 1864 when he was the conservator of forests in the central provinces of British India. There he had devised a system, adapted to local burning customs, of cleared and seasonally burned fire "traces" to surround protected forest zones, a scheme subsequently used throughout India.[10] In 1882, while Daniel Morris was transmitting his findings from Jamaica to London, Pearson's ideas and those of other international foresters were being taught by a combined German and French academic staff to young British foresters and others at the Ecole Nationale des Eaux et Forêts at Nancy.

In the early 1880s the new discipline of colonial tropical forestry was a shared, international enterprise under the unofficial sponsorship of the British Crown. Although the principal field observations and generalizations about tropical forestry came mainly from British India, the first British academic chair of forestry would not appear until 1887, and the first real training in imperial forestry in Britain did not start until 1905. Since the days of the British East India Company, however, practical problems of British governance in India had involved harvesting biotic resources from the subcontinent's vast and varied tropical environment, and British administrators there had developed a rich fund of empirical knowledge. Yet despite their administrative and managerial expertise in Indian forestry, the British in the 1860s had turned to a German academic, Sir Dietrich Brandis, and a corps of German-Austrian foresters. In India Brandis organized a regional forestry bureaucracy in the 1860s, and it was he who arranged for officers of the Indian Forest Service to be trained at Nancy in 1875 and for a school for Indian staff members in India itself three years later.[11]

Under the twin influences of Brandis's leadership and the expansion of the British Empire throughout the tropics, "Imperial forestry evolved into one of the mightiest institutions of colonial rule."[12] Its agents were scientifically trained and well-traveled field practitioners whose knowledge of the latest in tropical botany was augmented with such practical considerations as horticulture, land clearance, soil conservation, and *reboisement*. It was not unusual for men of the forestry service to have field experience in different tropical locales, applying what they had learned in, say, West Africa, to problems that arose in Burma. If they lacked direct field experience in particular places or specific problems, they could consult the growing fund of published knowledge (such as the tropical forest management issues about which Colonel Pearson had written) that was now being circulated worldwide by various London government offices. Nor was there always a sharp line dividing botanical/forestry interests and the sternly practical political concerns of colonial rule. By the late 1800s, British colonial governors and the lesser administrators who were the direct representatives of the London Colonial Office—men who usually had military backgrounds—occasionally had also written treatises on tropical botany, forestry, and related matters.

Daniel Morris of Jamaica was perhaps prototypical of this botanist/forester cadre serving the British Empire at the end of the nineteenth century. Educated in the natural sciences in Britain, his first overseas assignment dealt with investigating a coffee leaf blight in Ceylon in 1879. His posting as director of public gardens and plantations in Jamaica later that year began a distinguished agricultural career in the British Caribbean. In the next quarter-century Morris advised West Indian governors and planters on an informal as well as a formal basis as he traveled extensively, familiarizing himself with landscapes, crops, and personalities throughout the islands. A keen advocate of small-scale agriculture, Morris published many papers on its virtues as well as on the botanical characteristics and planting requirements of individual cash and subsistence crops. He was the London-appointed technical advisor to the West India Royal Commission formed in 1897 to investigate the plight of the sugar industry in the British Caribbean, and the following year he became the director of the Imperial Agricultural Department of the West Indies, which was established as a result of the commission. Upon his retirement in 1908, Morris became the scientific advisor to the colonial secretary in London on matters of tropical agriculture.[13]

Morris had few counterparts in the British Caribbean colonies, where despite the empirewide importance of the new science of forestry, the term "forestry" itself had an inappropriately odd ring. The reasons for this lack of fit were to be found in the singular ecological history of the region. Extensive forested areas—with the exception of British Honduras, British Guiana's interior, and a few of the mountain areas of Jamaica and the Windwards—had long since been removed to make way for plantations. Further, unlike in tropical Asia and Africa, there were few "natives" in the Caribbean whose traditional forest practices needed modernizing. Agricultural knowledge about the islands themselves was hardly an arcane or mysterious venture because it coincided with cash crop production and export, which always had been the British Caribbean's basis for existence. Except in a few isolated places, then, there never had been a need—at least in planters' minds—for Caribbean forestry. A formal forest service eventually was begun in Trinidad in 1901, after the deputy conservator of forests of the Indian Forest Department visited the island, "but in the majority of the British Lesser Antilles no forestry work had ever been carried out . . . on a scientific or comprehensive basis."[14]

So when Joseph Hooker suggested in late 1882 that the various British West Indian island colonies pay for a visiting forestry expert to travel through the region to render advice and assistance, not all reactions were positive. The Legislative Assembly of Dominica, for example, considered such a visit an unnecessary drain on local funds and reminiscent of the "worthless" visit to that island by Henry Prestoe of the Trinidad Botanical Garden in the mid-1870s.[15] In April 1875 Prestoe had come to Dominica to assess "the causes of the present depressed state of the Coffee cultivation, with a view to restore it to . . . its former importance." Although Prestoe was not, strictly speaking, a professional forester, he found in the interior of Dominica the same kinds of ecological wastefulness that Morris had written about in Jamaica. Of particular concern to Prestoe had been the destruction, by migratory highland dwellers, of coffee shade trees for short-term charcoal production.[16]

Despite a lack of enthusiasm in Dominica (and in Barbados and St. Kitts–Nevis) for sponsoring research and investigation by a visiting forester, the officials of most of the other islands expressed sufficient collective interest to support the travel and to obtain the services of E. D. M. Hooper, an official of the Indian Forest Department at Madras, to visit the British West Indies. Hooper arrived in the Caribbean colonies in late September 1885, and he departed in late July of the following year. During the ten months he was in the Caribbean, Hooper visited and assessed the forest conditions in Jamaica, Tobago, Grenada, St. Vincent, St. Lucia, Antigua, and British Honduras. He completed his summary reports about Jamaica and St. Vincent during 1886, and he finished his reports about the other places "in moments of leisure" during the next two years after he returned to his official post in southern India.[17] Each of Hooper's forestry reports was an assessment of the immediate environment itself as well as of the local institutions and laws intended to protect naturally forested areas. He found, especially in Jamaica and St. Vincent, the kinds of widespread and reckless forest destruction that Morris had reported a few years earlier. Hooper's principal overall recommendation, a point he raised in each report, was for a regional forestry officer to be assigned to the entire area, someone who would be able to coordinate the immediate action necessary to cope with local forest problems and to advise the several governments in the region.

In his written reports Hooper commented on and cited the latest thinking among Indian foresters and how these innovations might benefit Caribbean environmental preservation. On the subcontinent, imperial foresters adopted holistic views of their subject matter, yet their overriding concern was how to deal with fire. In the words of Berthold Ribbentrop, the inspector general of forests in India in 1900, "Fire-protection is the most difficult problem the Indian Forest Administration has to deal with."[18] Following this general concern, and reflecting his own identity and training as a forestry agent of the Crown, Hooper commented in each of his reports on the uses and misuses of fire in each of the Caribbean islands he visited. He found the "running fires" set by rural peoples to encourage new forage in the Antigua countryside potentially damaging to nearby plantations. In Jamaica he reported the forest destruction most devastating when the ground cover was burned after forest clearing. On the other hand, he commended the annual burning of high grasses along the country roads in Grenada, a technique proven effective in rural India. In most of the islands he advised more stringent legal controls against the damaging fires he found throughout the region, although he conceded in several cases that the laws that already existed were essentially "worthless," given the impossibility of identifying the culprits who had set the countryside fires in the first place.

Daniel Morris of Jamaica lauded the utility of Hooper's work and the value of his research and advice. On a visit to England in the spring of 1888, Morris delivered an address to the London Chamber of Commerce titled "The Vegetable Resources of the West Indies" (including within his remarks a lengthy exposition about the improper use of fire in the region). In his lecture Morris apprised his audience of the worthiness of Hooper's recent assessment of the Caribbean's forestry resources. Hooper's work could not fail "to have an appreciable effect upon the treatment and management of the forest still left in the West Indies." Morris also suggested that Hooper had "brought together a large amount of useful information respecting the condition of the interior of the islands, and the measures . . . necessary to their well-being and future prosperity."[19]

Some unofficial spokesmen were not nearly so kind and interpreted Hooper's brief visits as instances of meddlesome, and far too expensive, imperial bureaucracy imposed on individual West Indian colonies. In June 1886 the *Voice of St. Lucia* derided Hooper's brief visit to the island: "Mr. Hooper . . . arrived here on the 7th. . . . On the 8th he went out into the country exploring, and the weather he met with allowed of his immediately drawing up a very succinct report upon St. Lucia—'This island is all forest and rainfall—Received £150.' "[20] Two weeks later much the same reaction came from the *Antigua Standard*: "Mr. Hooper the Officer from the Indian Forest Department paid us a visit. He has been on top of the hills and down in the valleys, and returns in the mail to-day. We hear that his Report will be to following effect: 'Hills too low, not much good to be got by planting trees on them. Thanks. Received £200.' So much for Downing Street knowing more about what we wish than we know ourselves."[21]

These sour newspaper summaries of Hooper's visit likely typified the attitudes of business and commercial leaders, especially the planters, throughout the British West Indies. The severe economic depression that was creating the necessity for desperate measures and consuming so much of their time and attention had been caused, at least in their minds, by the unwillingness of the British government to apply restrictive tariffs to cheap European beet sugar. They needed help from London, but not in the form of an itinerant forester paid by local funds (already in very short supply) to take a cursory outsider's look at unproductive countrysides. The thousands of miles separating Hooper's permanent office in Madras, India, and meddlesome Downing Street was a geographical distinction that easily could be ignored by those who considered their own economic well-being jeopardized by the indifference of the officials of the mother country.

But the same economic depression also was inspiring fresh looks at wooded, non-sugarcane lands on West Indian islands as possible zones of alternative economic activities or as settlement areas for displaced and redundant laborers. Although the appointment of the regional forester recommended by the Hooper reports never came about, the subsequent posting of certain administrative officials to the region—men with genuine interests in their physical environments—was beginning to reorient local official thinking away from a total emphasis on sugarcane. Sir Cornelius Alfred Moloney, for example, was appointed governor of the Windward Islands in 1897. Moloney's military credentials included service in the Ashanti Wars in the 1870s, yet he also had published papers dealing with coffee production and other matters in tropical horticulture. In the last years of the nineteenth century, Moloney commented sensibly about the deforestation of Grenada's highland areas and suggested preventive measures that might be taken against it.[22]

Despite the differing environmental-historical trajectories among the various parts of the British Empire, a more efficient means of communicating knowledge of fire control and forestry from one place to another led to a certain standardization by the late 1800s. What Governor Moloney had learned in West Africa could be brought to bear on West Indian issues. The perspectives about controlling country fires brought from India by E. D. M. Hooper were actively reported and discussed, despite his detractors. The empire's civilizing mission thereby assumed a right-thinking environmental homogeneity. It of course assumed social control, and not only in urban areas. Flagrant violations of proper environmental stewardship, wherever they occurred, called for condemnation. A report in April 1894 of the so-called willful setting of fires in rural Tobago deplored the damage to property that these fires caused.[23] The description of these actions also implied that the perpetrators were intractable and impertinent. The visible and therefore public character of these illegal fires as insubordinate departures from prescribed orthodoxy—an orthodoxy buttressed not only by social proprieties but now by forestry's modern and scientific tenets—made attempts at fire suppression all the more desirable.

Forest Clearance, Environmental Conservation, and Human Depravity

Daniel Morris was by no means advocating a museumlike preservation of the Caribbean's remaining wooded areas as untouched, pristine forests when he condemned the wasteful and destructive burning of Jamaica's interior. Quite the opposite: he favored insular landscapes to be even more fully cultivated (but by using the right techniques) than they already were. In the same 1888 London presentation in which he reported on E. D. M. Hooper's visit to the Caribbean, Morris pointed out that Dominica's interior, as one example, needed badly to be developed. Third in size (after Jamaica and Trinidad) among the British West Indian islands, Dominica, according to Morris, required only "the right application of capital and energy to make it one of the most prosperous of our tropical dependencies." In its current "backward" position, however, the (lamentable) fact that Dominica was not a blooming tropical garden was "simply deplorable."[24]

The reorientation of backward wooded areas into production zones for tropical staple crops had, however, been accomplished many decades earlier for most of the British Caribbean islands. Only limited forested lands remained, and those usually were at inaccessible higher elevations. Perhaps the largest untouched area in the Lesser Antilles was the wooded lands of Trinidad east of the island's sugarcane belt. But by the 1880s small-scale cultivators and day laborers working under contract for large-scale planters were clearing and burning the island's central and eastern forest areas so that only the mountainous and hilly zones of eastern Trinidad remained covered by primary forest growth.[25]

Transportation improvements augmented by publicized exhortations accompanied and inspired the forest clearing. Trinidad's first railway line ran east from Port of Spain to Arima in 1876 and extended south to San Fernando in 1882, with a spur going up to Princes Town two years later. By 1897 the northern section of the line had been laid east to Sangre Grande, nearly all the way across the northern part of the island.[26] The railway sleepers were hewn from Trinidadian hardwoods, leading to some enthusiasm about the island's potential timber industry. More important to the island's economy, planters were now extending cacao cultivation into the central forests, an agricultural activity that might absorb excess labor from the depressed sugarcane estates. "Young Man, Go East!" exclaimed the *Port of Spain Gazette* in August 1889 in a lead article that enthused over the fecundity of Trinidadian forest soils, proclaimed an "awakening" of new agricultural possibilities, and reiterated an upbeat theme of development for the island that ran intermittently through the newspapers and official correspondence of the period.[27]

Projected rainforest clearing in the Windwards and the spasmodic construction of new roads and bridges there were more often attempts to reduce isolation and to bring the rural populaces under more effective government control than they were efforts to stimulate tropical horticulture. The estrangement between British rulers and the French Creole–speaking rural dwellers in both Dominica and St. Lucia was compounded by poor communications between the capital towns and the islands' forested interiors. A road tax levied on the men of both islands was an officially codified adaptation to the problems in deriving tax revenue from scattered populations of subsistence cultivators. The tough new Dominica Road Act of 1880, which penalized tax defaulters with one month's imprisonment at hard labor, sent police officials to rural villages to collect these taxes. When they arrived, however, they were told by the women that the men had emigrated to Venezuela. By the mid-1880s frustrated authorities and their representatives on Dominica are said to have collected local road taxes only "at the point of the bayonet," a strategy that did little to endear them to the rural peoples of the island.[28]

How could the problem of rural isolation be solved? "Roads, Roads, Roads," answered a newspaper editorial in the *Voice of St. Lucia* late in 1904. Better roads would help local authorities find out whether rural men had, in fact, departed for Panama and Cayenne. Better roads might even help to entice a genuinely loyal laboring class to populate the island. In the mid-1890s thousands of forested acres in St. Lucia were lying idle, "of no use to anybody." One suggested solution to the problems of inaccessibility and isolation was to import Barbadian laborers to build roads, clear the idle land, and populate St. Lucia's interior. Such a scheme would carry with it "the moral effect of anglicizing this British Fortress with English speaking men of enthusiastic loyalty."[29]

Whether Caribbean woodlands were cleared to provide cultivation acreage, improve intraisland accessibility, or satisfy imperial pronouncements, the organized removal by burning of a given area of virgin forest was a memorable event that involved several steps. Black axemen working under white supervisors felled individual trees into a "massive tangle." Then the trees' branches were "lopped and strewn" to blanket the cleared area, because an even cover of felled brush was as important as waiting (sometimes for several months) for the dry season for a "good burn." The initial fire sometimes was incomplete, so unburned logs and branches were piled at intervals in the charred clearings and then fired again. Care was obviously necessary so that managed burns stayed under control and did not spread to adjacent wooded areas and, even more important, to nearby grazing zones or to cacao and coffee plantings. The spectacle of the roaring fires of felled forest trees and dry brush that left stark and somber landscapes of blackened stumps surrounded by white and gray ashes held special meanings for the black woodsmen, some of whom believed the wooded island interiors to be haunts of jumbies and evil spirits who exacted vengeance on those who destroyed their forest sanctuaries. One way to avoid such retaliation was to have the forest clearance itself begun by a left-handed axeman.[30]

The long-term environmental effects of tropical forest clearing in general were, as they are today, imperfectly known yet the subject of considerable discussion. Although Daniel Morris had emphasized that Jamaican deforestation would lead to drought, his assertion was based more on conventional wisdom than on proven scientific fact, and it did not necessarily indicate that Morris equated tree removal directly with rainfall decline. Joseph Hooker, in commenting to the West Indian governors about Morris's memorandum, perhaps best captured the issue's ambivalence: "With regard to the influence of trees on actual rainfall . . . it would be easy to adduce arguments on both sides."[31] Probably the majority of foresters trained in India shared the stance taken by Berthold Ribbentrop. Ribbentrop asserted that the roots, branches, and leaves of forested areas served as "storehouses" for water, releasing water vapor directly into the atmosphere through evaporation and thereby contributing directly to rainfall in a given region.[32] Data from France and Germany presented as evidence in a forestry lecture in Port of Spain in March 1903 supported the idea that more forests meant more rain, although these comparative data had come from sites at different elevations.[33]

In the West Indies, ideas concerning the relationships between deforestation, burning, and rainfall were aired with far more conviction than scientific certainty. An anonymous letter to the Antigua newspaper in January 1889 acknowledged that the idea of trees encouraging rainfall had long been "accepted doctrine" but that "exact investigations in the States, in Europe, and in India prove that this is *not* the case."[34] In countering the Antigua letter and similar pronouncements, the regionwide royal commission nearly a decade later condemned the practice of squatting on Trinidad forest lands because the clearing and burning of forest plots "would injuriously affect the rainfall."[35]

But the real value of forest covers lay not in whether they directly caused rain to fall. The shade and physical protection that forests provided insular soils was a more obvious, subtle, and practical issue, relevant to both large-scale forest removal and small-scale clearing. Daniel Morris lectured to a group of Jamaican planters in 1885 as to how deforestation worsened the effects of climatic extremes, a point he tirelessly reiterated elsewhere in the region for the next twenty years.[36] Similarly, E. D. M. Hooper's written reports about both St. Lucia (in which he said the idea that forests actually attracted rain was "out of date") and Grenada emphasized that a proper vegetation cover conserved ground moisture, and his reports were cited approvingly as both authoritative and sound in the latter island into the

1890s.[37] In February 1889 the *Port of Spain Gazette* adopted a somewhat apologetic tone—apparently to mollify development enthusiasts—in editorializing about the practical need to preserve wooded areas near the rivers, springs, and reservoirs that provided the island's water supply. It was crucial to recognize, according to the article, that a forest's "primary function is that of a great vegetable sponge that gathers moisture from the air and clouds, and stores it up for distribution."[38] A decade later Sir Cuthbert Quilter of London's West India Committee communicated directly with the secretary of state for the colonies about the destructive soil erosion produced by deforestation in Trinidad, in particular how it had led to the silting of the Caroni River in the central part of the island's sugar belt.[39]

These and similar warnings, combined with decreasing sugar prices, led to a certain local interest in conservation. Among some observers in Grenada, Hooper's recommendation for a conservation law to protect forested mountaintops took on an urgent tone. The recent and widespread cacao planting on previously forested slopes there had led to muddy sheet erosion and flooding in the rainy seasons. During the other half of the year, in Grenada's dry months, many of the island's streams had "shown signs of drying up," a seasonal problem that, by 1887, had concerned observers "calling for a compulsory law for forest preservation . . . enforced by an Inspector of Forests."[40] Conservation laws such as these were not altogether new in the islands. A 1721 ordinance in Antigua that protected trees near the island's Body Ponds (a series of natural reservoirs) was still in force in the late 1800s. The King's Hill Forest Act of 1791 in St. George parish on St. Vincent continued to protect, at least in theory, a small forest reserve "for the purpose of attracting the clouds and rain."[41]

A botanic station in Trinidad had existed, publishing annual reports about local agriculture and crop experiments, since 1872. A quarter-century later, and as part of a celebration of a century of British rule in the late 1890s, some Trinidadians attributed what they considered a sensitive and thoughtful environmental awareness on their island to their regionally unique Spanish heritage. Although British Trinidad had no formal Forest Department (as in some Asian parts of the empire) until 1901, its Crown Lands Office housed local forest inventories and even instructions for woodland preservation that were said to be minutely detailed and the work "of the former Spanish Governors."[42] Elsewhere in the region by the turn of the century, and again under the leadership of Daniel Morris, nearby islands had established government-sponsored botanic agencies whose resident botanists pushed for at least minimal environmental protection. St. Vincent resuscitated its once-famous botanic garden in 1890, and new botanic gardens or botanic stations were also begun in Grenada (1886), St. Lucia (1887), Antigua (1889), Dominica (1891), St. Kitts (1899), and Montserrat (1900). While it would be tempting to suggest that these gardens and their resident botanists represented the well-defined contours of a new environmentalism that began to extend its influence throughout the region, these agencies and their personnel just as often were ridiculed as wasteful extravagances. Shortly after the new botanic garden was started in Grenada, for example, the curator was derided as worthless: "Mr. Elliott, for all the good he is to Grenada, may have been Curator of a Botanic Garden in the Moon."[43]

Whereas reports and sentiments were mixed about the efficacy of government botanic offices to promote environmental protection, near-unanimity characterized the published condemnations of destruction of local forests by fire-wielding, small-scale cultivators. Passionate concerns about forest destruction and the general official dislike of nomadic upland dwellers inevitably led Daniel Morris and others to brand them as inferior human beings and worse.

When Henry Prestoe visited Dominica from Trinidad in the mid-1870s, his shock at what he considered environmental abuse was overshadowed only by his disgust for the itinerant forest dwellers themselves. He found "this depraved class of squatters" particularly numerous in the "hills in the north of the Island, where they have located themselves most distantly from civilizing agencies." Prestoe did not specify ethnic identity, but it is probable that those in question were mainly of African descent, as he described them as having "relapsed" (the usual term indicating that local blacks had reverted to primitive African ways) "to a state of semi-barbarism." These people spoke the most debased form of Creole Prestoe had ever heard and dressed their children in colorless rags. Their vile, degraded material culture consisted of a few broken plates and pots plus some half-starved dogs that slunk around the campfires. The thatched, earthen-floor dwellings they inhabited, moreover, seemed "inferior in all respects to the sheds which the Coolies in Trinidad put up to protect their cattle." In sum, the entire existence of these peoples was "in a moral sense . . . deplorable."[44]

John Hart, the superintendent of the Royal Botanic Gardens in Trinidad, told similar tales about those responsible for the widespread "felling and burning" in the eastern part of the island in 1891. This wanton destruction was accomplished "by a roving set of semi-civilized African squatters, principally 'Congos' and 'Cangas,' who, after reaping one or two crops of rice or corn, abandoned the place and wandered to some other convenient locality."[45] Hart's condemnation of "African" forest destruction, typical of the widespread remarks about African savagery and wickedness in the contexts of ceremonial and malicious fires in the towns and villages of the region, represented a colonial mind-set that became hardened into regional economic policy by the end of the decade. The 1897 Royal Commission (and much of the official correspondence generated by its suggestions) strongly recommended small cultivator plots in the region and emphasized the need for such plots to be located near towns and roads so that unsupervised local cultivators would not regress into African modes of production.[46]

In a widespread effort in condemnation by newspaper editors and government officers that usually took on a righteous, save-the-environment tone, special official denunciation was reserved for the producers and purveyors of charcoal. Charcoal, along with firewood, was used for almost all of the cooking in the islands; it therefore represented—on a locally tiny but universally widespread scale—the most fundamental and common fuel in the eastern Caribbean in the late 1800s. Working peoples rarely stored charcoal because of dampness and lack of space, but mainly because their tiny incomes allowed them to buy only small amounts at a time. Usually they purchased it daily at village and city markets. Vegetables, fish, and imported foodstuffs all had their seasons, but the most common items of every marketplace, regardless of the time of year, were small piles of charcoal priced at a penny apiece. Every day, patrons purchased "about as much as would serve to boil a tea-kettle. . . . Nobody bought more than a cent's worth."[47]

One could reasonably argue that charcoal was the most common trade item in the region at the turn of the century. Yet it rarely, if ever, appeared in import-export tables of the time because both production and marketing were ubiquitous and usually at a very small scale. On most islands it was a mountainside-to-market commodity; men brought bags of charcoal to daily markets and parceled out the contents to individual market women. A robust, if shadowy, interisland schooner traffic in charcoal also satisfied the demand on islands where local production was limited. Both Grenada and Tobago, for example, shipped charcoal to Barbados on the decks of small sailing vessels. But charcoal making also was common, and perhaps more ecologically devastating, in the arid northerly places. In 1893 an official on Anguilla lamented that some native species had been cut over two and three times because locals "cut all timber to burn for charcoal."[48] On Tortola in 1897, charcoal making and exporting represented "a very important source of income." The naturally shallow soils of Tortola were "laid bare," and charcoal production there was directed toward the market in Danish St. Thomas. Small sailing sloops each carried forty to fifty barrels of charcoal at a time from Tortola to St. Thomas, a collective quantity that had been produced "at one burning" on the former island and then allowed into the latter via a "system of fraud in the declaration of these loads of charcoal."[49]

Insular colonial officials, only vaguely concerned about the ill effects of charcoal on the environment and probably far more cognizant of its importance to local economies, did their best to monitor and regulate production and marketing. But controlling charcoal making and selling was as impossible as regulating the distillation and consumption of illicit rum. Police nonetheless harassed and hounded charcoal sellers who had not bought licenses or paid the proper marketing fees. In February 1885 on the wharf in Bridgetown, Barbados, police "raided" the purveyors of firewood and charcoal for selling these commodities destined for local stores.[50] In 1901 in St. John's, Antigua, they denied the right for charcoal venders to sell door-to-door and required them to take their merchandise to the local, government-controlled marketplace, thereby hindering these peddlers' attempts "to earn an honest living."[51]

Charcoal license laws, where they did exist, bore little relation to the local reduction and processing of native timber into small chunks of cooking fuel. Although 1,275 charcoal licenses were issued in Trinidad in 1890, the law requiring them was considered a "dead letter" as far as forest protection was concerned.[52] Especially on St. Vincent, where in 1886 E. D. M. Hooper noted that charcoal makers had reduced the slopes of Mount St. Andrew in the far southern part of the island to "complete barrenness," charcoal legislation had a lengthy, if ecologically incongruous, pedigree.[53] A charcoal ordinance had existed on St. Vincent as early as 1839, and the law in force when Hooper visited had been enacted in 1844. A new charcoal ordinance of 1888 was amended in 1890 and then again in 1891, although in 1897 its provisions continued to delineate a tedious, unwieldy, and expensive regimen for anyone on the island who wanted to (legally) make and sell charcoal. District magistrates were to issue annual charcoal licenses for a fee of 2 shillings, 6 pence, after they were satisfied that the applicant had approved access to the land in question. Next, the charcoal producer was to obtain, free of charge, a three-day removal permit from the police station. A "general" license costing 5 shillings allowed one to peddle the charcoal, and further licenses were necessary to sell it in shops in Kingstown (20 shillings) or in rural areas (10 shillings). In 1897 the island's administrator, in a case of classic understatement, pronounced these unrealistic regulations "vexatious" and noted that "the result obtained from them does not justify their continuance."[54]

Others questioned even the morality of this legislation and suggested that the high fees for charcoal production and marketing exemplified laws designed to extend the historic "oppression of the working class."[55] As true as this comment apparently was, it did not specify the probable real reason why these unrealistic regulations persisted. Despite the commentary as to the lack of fit between law and reality in matters of charcoal making and selling, the St. Vincent Legislative Council upheld the letter of the law.

According to its key proviso, to ensure that the many sequential fees were paid and steps taken to satisfy legal charcoal activities, the police had unlimited "powers of search and seizure if charcoal is found or suspected to be in a person's possession."[56] The island's officials and planters thereby maintained a legal means of surveillance and authority over the woodsmen and itinerant forest dwellers on St. Vincent whose remote locations normally placed them outside the daily control of lowland officials.

Legal issues aside, charcoal making seems to have placed serious ecological stress on forest environments. The simple technology and clandestine forest locations help to explain why. Several men with axes and shovels could dig a pit, fill it with chopped wood, and then smolder the logs and branches inside the large covered hole—where the oxygen supply was limited—thereby producing charcoal for sale within a few days. Afterward these and other men were willing to risk encounters with the few forest rangers paid by planters and local governments to monitor insular woodlands. The cumulative results were the cutover hillsides noted by Hooper and a number of other officials. As early as 1880 the vegetation on the slopes of the King's Hill area in St. Vincent, the prototypical conservation site in the southern Caribbean, was being chopped down by a variety of licensed and unlicensed charcoal makers, leading to high rates of slope erosion.[57]

The widespread regional demand for charcoal meant that the hillside slopes of often the tiniest islands in the region were laid bare to supply markets on other islands. Not only were Anguilla and Tortola experiencing environmental reduction in order to satisfy charcoal users in St. Thomas, but small islands in the south responded to Barbadian charcoal demands in the same way. Carriacou and the St. Vincent Grenadines, which were only occasionally monitored by local officials, supplied sloopIoads of charcoal for the Bridgetown market, to the detriment of their own insular habitats. One forestry official in 1889 suggested that Tobago's environmental devastation, accomplished to satisfy Barbadian charcoal demand, was the result not of too much money being offered for the product, but too little. The thin profit margins realized by charcoal exporters from Tobago to Barbados had forced them to seek out smaller and smaller plants, in effect driving their own environment harder and harder and burning it more often to meet external market demands.[58]

"The Harmfulness of Bush Fires"

The general term "bush fire" was always applied to the nonagricultural fires occurring in the countryside outside cities, villages, and plantations of the Lesser Antilles. Although nearly all were anthropogenic fires—such as accidental hillside fires sparked by embers from torches or "firesticks" used by night travelers in the Windwards, or the more deliberately set seasonal blazes ignited by herders in all of the islands—their frequency and extent were closely related to environmental factors. During an abnormally dry May and June in Trinidad in 1889, for example, a fire smoldered for a month in the vicinity of the normally wet Oropouche Lagoon southeast of San Fernando before it burst into full flame and swept over roughly 120 acres, destroying provision crops and singeing hundreds of coconut trees.[59] One decade later, after the hurricane of 1899 devastated several of the Leewards on its way to Puerto Rico, hillside burning on Montserrat compounded the damage caused by this natural hazard. Montserratian land renters burned off the trees felled by the huge storm in one mountainside area, leaving nothing to hinder the severe erosion caused by subsequent rainstorms.[60] Only on rare occasions, such as when lightning ignited trees in Barbados in October 1890, were bush fires begun by strictly natural causes.[61]

Drought was the natural ally of human-set bush fires. During periods of aridity, whether on the lower islands such as Antigua and Barbados, where insufficient relief often led to a paucity of mountain-triggered rain, or in the entire region when high atmospheric pressure prevailed for long periods, dry conditions were invariably accompanied by successions of bush fires. During the 1890s, in normally wet Trinidad several dry and windy periods led to alarming rural fires that threatened nearby settlements and crops. Unusual dryness there beginning in December 1890 led to forest fires in the ensuing months that local agencies could not control. In April 1891 one of those fires burned over much of the mountainous catchment area that intercepted rainfall and eventually channeled it into Port of Spain's water supply. In mid-May these fires were burning in full view of the Trinidad governor's residence.[62] Similar conditions there in 1898 and 1899 led to fines levied on individuals who lit fires without licenses. In late May of 1898 one firsthand report from Trinidad told of how widespread these rural fires had become: "All over the country are to be seen every night numbers of bush and hill fires, some of very large dimensions. By day also very large fires are frequently set, and in not a few instances they have to use a native form of expression, 'got away.'"[63]

As in other times and places, physical conditions conducive to burning inspired and magnified human propensities for setting fires. During an unusually dry first few months of 1901 in St. Lucia, a number of bush fires plagued the rural areas of the island. Unable to control either the atmospheric or the human conditions that encouraged the starting of fires, local authorities developed a rough, tripartite typology of rural bush fires. The first category was the accidental fire, often caused by children playing with matches; the second involved criminal intent; and the third was the "good faith" fire, usually small brush-clearing blazes that had gotten out of hand. To reduce the number of fires, especially those that fell into category two, government officials in Castries sought help from the French-speaking Catholic clergy in the country districts to exercise "moral persuasion."[64]

Frustrated because of their inability to control local environmental events and naturally disparaging toward the people who chopped down local forests for subsistence cultivation, British officials in the Lesser Antilles tirelessly condemned bush fires and those responsible for setting them. How could dry Antigua, for example, ever achieve reforestation if local herders of sheep and cattle heedlessly burned over the range, thereby destroying tree seedlings struggling for survival? Francis Watts, the government analytical and agricultural chemist for the Leeward Islands, raised these issues in a paper in 1900. His article dealt with the value of reintroducing a forest cover to the island, and he raised the issue of careless burning at the end: "I am compelled to introduce it [fire] here because of the disastrous effects upon seedling trees and young timber which these uncontrolled fires have, sweeping across the country and burning for days together; but for this it seems probable that much of the island would be covered with heavier timber. I introduce it to condemn it."[65] Suggested reforestation proposals (probably little more than wishful thinking on the part of government agents) in St. Kitts and Montserrat were similarly doomed to failure because during the dry seasons rural livestock owners always lit grass fires that swept through the dry brush of the cleared lower reaches of mountainsides until they encountered the dense and moist stands of subtropical forest.[66] Those who lit the fires, furthermore, were almost never caught. So the official condemnations of bush fires perhaps reflected as much frustration over an inability to identify and apprehend rural fire starters as they did anything else.

The small-scale herders in the Lesser Antilles left no written records, obviously, of their annual burning practices, which were techniques passed by word of mouth from one generation to another and from neighbor to neighbor. But the few written records that describe their activities provide hints that, through trial and error and with little more than their own experiences to guide them, these individuals, through the judicious use of fire, routinely produced forage each year for their grazing stock without irreparably harming their immediate environments. Observers of these burning practices usually used the term "running fires" to characterize the common, low-intensity clearing fires that these pastoralists employed to burn grasslands. The descriptions of these widespread, omnipresent fires along with the accounts of the willful stubbornness of those who insisted on lighting them suggest a practice repeated time and again. Without pushing thin evidence too far or overly romanticizing peasant wisdom, we can surmise that these rural livestock keepers were likely practicing sound range management techniques that have subsequently been endorsed in modern ecological thinking. Among other enhancements, controlled burning of pasture lands today is often thought to improve subsequent plant productivity by warming the soil, providing plant nutrients in the form of ash, and reducing aging and competing shoots and litter.[67]

The possibility that Antillean range firing practices might be doing some good did not go entirely unnoticed. In 1901 William Fawcett, the director of public gardens and plantations in Jamaica, enjoined a (mainly condemnatory) discussion about the widespread practices of rural burning in the West Indies. Fawcett told of how peasants along the slopes of the Blue Mountains in Jamaica burned the grasslands each year to improve pasturage, a practice so widespread that local antiburning laws were of little use. Fawcett further pointed out that official condemnation against burning was perhaps being taken too far, especially by those who advocated a ban on fires altogether. He supported his assertion by reminding colleagues that "on certain clay soils in England, the grass is cut and burnt in heaps and the texture of the soil is thereby improved."[68]

Rural fire could not be banned entirely, nearly all agreed. It was necessary in forest clearing because there was simply no other way to eliminate unwanted brush and timber. Further, overall economic changes in the region could perhaps mean a greater appreciation for controlled forest burning. With the low sugar prices creating such hardship, the innumerable exhortations from many sources about alternative forest crops that could augment or even replace sugarcane as the region's main cash crop were understandable. Clearing for such crops often meant burning forests and brush on mountainsides, especially in the volcanic Windwards. One observer in St. Lucia in 1901, recalling his fifteen years' experience, asserted that "it is absolutely necessary to burn forest land if cacao, coffee, limes or ginger are to be cultivated."[69] Even after cacao was planted in Grenada at the turn of the century, burning was considered part of the cultivation routine. In the island's upland cacao plots, many of the small-scale planters had seasonally burned refuse and old leaves for two decades, as it took too long for this material to rot in place. One drawback to this practice was that cacao litter and leaf fires, especially in the mountains' infrequent dry periods, often escaped and scorched nearby trees.[70]

In rare cases, long-standing countryside fire practices were incorporated into official campaigns to eradicate insect pests. It was well known that as far back as anyone could recall, rural cultivators had built fires to reduce—by the use of fire and smoke—crop pest damage and general insect an-

Bush fires, such as in Nevis in 1976, have always cleared brush and reduced insects in the Lesser Antilles (photograph by the author).

noyance, a measure that brought, at best, only temporary relief. And smoky fires were common preventive tactics used by nearly everyone against mosquitoes. During the latter half of 1885, reports from northern Venezuela about local agricultural damage caused by swarms of locusts were monitored closely in Trinidad. Late in the year the insects appeared in the southern part of the island as well as at Chacachacare, the most westerly of the minuscule islets off Trinidad's northwest coast. At the latter place, steps were taken to confine the locusts to one area of land and then to eliminate them by circling "the whole place, which is bare ground, by fire. . . . No damage has been done whatever to either coconuts or cocoa."[71]

If one could return to any of the British Antilles at the turn of the century and query colonial agricultural officers about the local effects and overall regional importance of local bush fires, almost certainly one would be directed to the contemporary writings of an influential resident of Dominica, Dr. H. A. Alford Nicholls. Nicholls, a medical doctor, planter, public figure, and author, was also a corresponding member of the New York Academy of Sciences with similar relationships to the agricultural societies in Guadeloupe and British Guiana. Like other leading planters and businessmen of the region, he also traveled on occasion to nearby islands, where he consulted with local planters and agricultural authorities and made his own observations. Nicholls was particularly concerned with the environmental damage caused by large-scale bush fires. Among his more general writings about tropical agriculture, he described the damage by the widespread fires he observed on Dominica, explained why he considered them ruinous, suggested preventive measures, and extended his generalizations to include the other British islands in the region.

Nicholls was by no means a cantankerous pyrophobe whose social standing provided an outlet for his rantings. In the late 1880s he submitted to the government of Jamaica the prizewinning manuscript in a competition for a suitable school textbook on tropical agriculture. *A Text-Book of Tropical Agriculture* was then published in Kingston in 1891 and in London and New York the following year. Nicholls's book described overall environmental characteristics of the tropics, included sections on climates and soils, identified appropriate cultivation techniques and agricultural implements, and surveyed a wide variety of commercial crops. Nicholls illustrated the book with his own line drawings as well as diagrams available from Kew Gardens.[72] He dedicated the text to Joseph Hooker because of the encouragement and advice Hooker had provided while Nicholls had been "the Kew correspondent for Dominica." Nicholls's book, along with *Blackie's Tropical Readers*, was a standard written text used by agriculturalists, including "pupil-teachers," in the British Caribbean during the 1890s. It is most likely that, as a result, his advice in tropical agricultural matters—including his suggestions about burning—was considered highly authoritative in the Lesser Antilles and elsewhere.[73]

In August 1899 Nicholls presented an antifire resolution to his fellow members of the Dominica Legislative Council for approval. The measure was intended to empower the governor of the Leewards to prohibit open fires in times of drought. The resolution passed unanimously on the basis of Nicholls's presentation. He pointed out that "year after year, during the dry season, planters . . . suffered great losses by fires set by their neighbours." He recently had discussed Dominica fires with planter colleagues who reported the damage done to cultivated and uncultivated lands by the widespread, human-set fires on both the windward and leeward sides of the island. Nicholls pointed out that these fires destroyed nitrogen-rich litter that, if allowed to decompose, would increase fertility and that burning "nitrifying microbes" in the upper layers of the soil also was environmentally injurious. Nicholls expressed concern about the fire-reduced inorganic components of the soil (especially on the dry leeward side of Dominica) and the subsequent erosion because of it. The worst damage was in the relatively low, potentially drought-prone coastal zones of the island, as the high areas "are still covered with the primeval forest."[74]

In the following year, worried that the government of Dominica had not yet passed into law the antifire resolution of the local legislative council, Nicholls developed and presented a detailed typology of West Indian bush fires. It was based on his and others' observations on Dominica and elsewhere. The typology appeared in the Dominica and Trinidad newspapers (and was titled "The Harmfulness of Bush Fires" in the latter), and it was summarized and published as an article as well as a circular.[75] In any case, Nicholls argued that bush fires were bad enough on Dominica, but given the heightened concern about rural fire throughout the region, they constituted an interisland menace. Bush fires also could become sugarcane fires, and vice versa, a point with obviously important economic implications for the whole region. The circular letter from Malcolm Kearton & Company of London warning of reduced insurance coverage because of increased rural incendiarism had, after all, appeared in several West Indian newspapers in 1898. And Barbados was suffering an epidemic of sugarcane fires in every growing season of the late 1890s. Beyond urging local antifire actions, Nicholls also was in direct contact about the problem with the West India Committee in London.

The typology itself identified five classes of rural bush fires. The first was the burning of diseased plants ("one of the heroic remedies of the plant physician"), which was routine among cultivators but that, according to Nicholls, should never be undertaken without strict precautions to prevent more damage than necessary. The second was the anti-insect fire, the widespread practice used to smoke out harmful pests. The third class of bush fires was the well-known "burn" used in organized forest clearing, the steps for which had originally been outlined by "Laborie, in his well-known work entitled *The Coffee Planter of Saint Domingo* published in 1797." The fourth was the seasonal grass fire that local herders set to induce young shoots for better grazing; these blazes commonly escaped, according to Nicholls, to create more harm than good. The last of the five classes of rural fires was the "ordinary and well-known bush fires of the tropics," employed to clear land prior to cultivation, a practice deplored by Nicholls mainly because of the reduction, through burning, of organic nutrients in the soil.[76]

Nicholls did not develop his bush fire typology as an academic exercise. Rather, he hoped it would provide evidence to encourage decision makers in Dominica and throughout the region to initiate antifire legislation. He found a somewhat receptive audience for these ideas at the regional agricultural conference in Barbados in 1901. Among their other activities, some of the agricultural officers discussed his ideas, mused about differences and similarities in their own particular islands, and pointed out that many antifire laws already existed, not necessarily as bush fire edicts but within various forest preservation acts. The group also perused representative antiburning laws from Jamaica, Mauritius, Cyprus, Ceylon, and Trinidad.[77]

Nicholls might have made an even stronger case had he reminded his fellow conferees that bush fire legislation in the region usually had originated immediately after disastrous fires that easily could recur. Such a reminder had been deemed necessary in Trinidad in May 1899, when Mr. St. Hill proposed, at a meeting of the agricultural society, a relaxation in the local edicts against burning farm refuse in the dry season. The *Port of Spain Gazette* provided a sobering rejoinder to this proposal, an answer steeped in history yet vivid in capturing a sense of what still could happen at any time given the wrong conditions:

> The dry season of 1869, like the present one, was very long, and just as is now the case, was marked by strong easterly winds. Fire after fire broke out in the hills and at one time the whole Northern Range was a-blaze from the Bocas to Blanchisseuse and Toco. On one occasion the fire swept down from Maraval and so seriously threatened the residence of the Governor. . . . Another time the fire broke out close to the Anglican Chapel at La Ventille, and spreading with fearful rapidity before the strong breeze from the East, approached in dangerous proximity to the Powder Magazine. . . . Night and day the Police in the various country districts were on duty and large bodies of labourers had to be employed for the purpose of fighting the enemy then advancing in all directions. . . . For weeks the whole atmosphere was hazy with smoke and the air filled with black particles of burnt leaves. There was naturally a general feeling of alarm and the nights were passed with the greatest anxiety, no one knowing where next the fire might not break out. It was seen that some legislation was needed . . . and consequently the Ordinance No. 24 of 1869 "For the prevention of Accidents by Fire" was passed.[78]

Whereas the bush fire threats on Trinidad and other islands were, obviously, of the utmost importance to the people residing in those places, Nicholls hoped to organize support for regionwide fire prohibition legislation coordinated by the various colonial governors. At the 1901 conference in Barbados, as he had during the preceding year, he asserted that he had been in contact with officials at the Colonial Office and that, in antifire matters, "the Secretary of State wants uniform legislation throughout the islands." Yet the problems in such a strategy, according to Nicholls, were the bedeviling differences among the various places in the region, an area that the Colonial Office realized was composed of "islands differing widely in physical features, in industries, in the character of the people and even in language."[79]

Although Nicholls cited no examples beyond the Caribbean, it is possible that officials in London were aware of the problems in trying to introduce homogeneous antifire legislation in such a varied region. After all, they had encountered such difficulties in India, the laboratory and proving ground for modern tropical forestry. Whereas the Indian subcontinent was not a series of discrete insular colonies, its latitudinal and topographical differentiation made standardization in fire control nearly impossible. Any official trying to compare fire protection in India for even two forested areas had innumerable factors to consider, including "the number of dewless nights, the force of the wind, whether variable or constant, and the hours of the day or night that it blows; the facilities for procuring assistance, the presence or absence of forest villages, the existence of rights of grazing, of collecting minor produce; the existence of main highways of traffic through or near the protected area . . . and many others which will occur to any forest officer."[80]

The maddeningly fractious particularism that British colonial officials encountered at the turn of the century in the many settings and circumstances of the Indian subcontinent and the varied landscapes of the insular colonies of the Caribbean became obvious mainly to those on the ground. Homogeneity and order might be conceived of and planned in London or even in an office in one of the capital cities of the Caribbean itself. But outside the towns and cities the varied terrain, seasons, economic activities, and even personal whims posed and presented a variety of problems and opportunities for rural dwellers. A number of these problems and opportunities could be addressed by lighting fires of varying and changing sizes and intensities, depending on many variables. Colonial officials, in their quest to classify, grouped all of these blazes as "bush fires," a definition based on proximity as much as anything else. A bush fire was outside their domain, an annoying event that normally posed little threat to settlements or property and which usually could be ignored. That was best in most cases because any attempt to identify, much less apprehend, those responsible was sure to end in frustration and failure. Certainly the burns that accompanied planned and officially orchestrated land clearance projects were necessary and comprehensible. Beyond them, there probably were as many reasons for lighting rural fires—H. A. A. Nicholls's classification scheme notwithstanding—as there were people igniting them. Probably realizing this diversity in fire motivation and rural fire types, British Caribbean colonial officials must have learned to live with bush fires unless they encroached on towns and cities. But when rural fires threatened the economic lifeblood of the entire region, the local sugarcane crops, official indifference was no option.

5

Sugarcane Fires

A CARIBBEAN SUGARCANE FIRE IS OFTEN A spectacular event. It is always intensely ecological because it is at once an environmental, biological, and social occurrence. A sizable cane fire is momentous for a small island because it attracts nearly everyone's attention, and if it spreads to adjacent fields and wastelands, as it often does, it thereby affects directly a considerable portion of the place. News about the whereabouts, sizes, directions of travel, and possible origins of cane fires travels quickly and is often embellished with each retelling, so one need not see a sugarcane fire to get caught up in its excitement. A sugarcane fire also generates an often irresistible temptation to start other cane fires nearby; so these fires sometimes come in bunches or clusters, and groups of obviously related sugarcane fires have occurred in the Caribbean islands for many decades. It is not too much to say that the presence of a cane fire, or a series of cane fires, has the contagious capacity to energize an entire island; everything and everyone seem drawn to and focused on the flames and smoke. This coalescence of people and energy and excitement can last momentarily or for several days, or it can help to animate a Caribbean island's human society for an entire harvesting season.

Seasonal and climatic conditions greatly influence sugarcane fires.[1] Young, tough, fibrous, green canes are very difficult to burn, especially on a still morning when the air's relative humidity is high and the dew is still on the leaves. At the other extreme, a field of mature sugarcane ready for harvest at the end of the dry season can fairly explode into flame when ignited during a late afternoon. When the wind or sea breeze is high, as it often is in the afternoons, a sugarcane fire spreads quickly, jumping firebreaks, ditches, and paved roads and sometimes inflicting severe damage and crop loss, creating conditions that present "probably the greatest risk to which a sugar-cane plantation is exposed."[2] When Jamaican-born historian Franklin Knight, in describing vividly the infectious discontent that sparked the pivotal social disturbances of the late 1930s in the British Caribbean, declared that "labor unrest raced through the Caribbean like fire on a windy day," he has to have had a West Indian sugarcane fire in mind.[3]

There are different kinds of sugarcane fires and different ways these fires can be classified. At the beginning of the twenty-first century, although sugarcane is beginning to disappear from many of the islands, probably most of the cane harvested in the Caribbean region is purposely burned prior to cutting. This intentional burning is far different from the incendiarism of one century ago, when any cane fire was normally considered an illegal act of arson. Today preharvest firings are sometimes known as "cool burns," backfires that are monitored and well controlled and that are mainly leaf and trash fires that barely singe the standing stalks of cane. But any fire can become hotter, even when the air is still. Veterans of many cane harvests occasionally speak of a cane fire "chimney," or the recognizable center of a fire from which ash and heat waves ascend into the sky. These fire centers, because they are tiny, focused, and upwardly spiraling vortices of low pressure, generate surface breezes across the fields into themselves. These breezes provide more oxygen to the fire and are also thought by some to cool and insulate the soil, thereby protecting it from baking from the fire. But most Caribbean cane growers agree that the possible environmental harm of cane burning differs from one part of the region to another, being less dangerous on the heavy clays of Guyana and Trinidad than it is on lighter volcanic soils such as in St. Kitts and St. Vincent.

During the daytime, columns of smoke—visible from several miles away—indicate the cane fire's location. Sometimes the smoke appears black and oily, like that from an industrial fire. Damp sugarcane that is burned usually gives off white smoke, with billows of water vapor and nitrogen escaping into the atmosphere. On sunny days, thick clouds of smoke from a cane fire cast moving shadows that darken the fields momentarily, depending on the direction of the prevailing wind. As you get closer to the fire, the translucence of the ambient smoke turns everything orange-gray. The smoke from cane fires produces traffic hazards for motor vehicles and pedestrians alike. Especially on paved roads during the sugarcane harvest, it is not unknown for large trucks laden with cut sugarcanes to hurtle headlong through clouds of smoke en route to the mill and even through smoky road intersections, their drivers paying little attention to their hazardous and reckless behavior. The loud crackling noises that a sugarcane fire emits are particularly noticeable at night. After dark the flames are visible for long distances, and at a closer range the orange and yellow flames appear very dangerous. The most striking historical narratives about the terror and emotion evoked by Caribbean sugarcane fires seem always to dwell on descriptions of those fires occurring at night.

The surge of heat given off by a Caribbean sugarcane fire is not unlike that from a closed oven when you open the door. The heat and flames from the burning fields send rats and mongooses scurrying in all directions. Clouds of hornets and other insects that have nested in the fields during the crop's growth fly away. Snakes try to escape the flames. In Barbados, large light-brown centipedes with pink legs angle their way across paved roads and seek refuge under and inside houses adjacent to the sugarcane fields. The planned, preharvest fires are usually explained to visitors as necessary to eliminate the rodents, snakes, and insects, as well as the sharp sugarcane leaves, that inhibit the arduous work of manual canecutting.

But the cane fires have further, related side effects. Birds swirl through the air, energized by the heat and flames. White and gray egrets appear at the edges of the fires to catch the escaping insects; these large and graceful birds parry and thrust with the edge of the fire, taking care not to be burned. In the intermediate slopes and volcanic highlands of St. Kitts, small-scale subsistence farmers complain that sugarcane fires send the insects to devour the food crops in their own fields, an ecological dimension of the enduring asymmetry between large-scale and small-scale agriculture in the Caribbean region. Wisps of black ash from sugarcane fires travel for miles, soiling damp laundry that has been hung out to dry or sailing through the wooden lattices in village windows, reminding everyone—as if they needed reminding—that "the crop" is under way.

A sugarcane fire also touches off a chain of interlocking economic activities. Harvesting must follow quickly because, after burning, the cane's sucrose content begins to fall, and it declines rapidly after about twenty-four hours.[4] The burned cane's deterioration is particularly acute if it becomes wet before harvesting and milling.[5] So when the canes from more than one field or more than one estate or holding are destined for a single grinding mill and planned burning precedes harvest, the sequential timing and coordination of sugarcane fires are crucial and often complicated. That is why unplanned, "malicious" fires pose monumental problems for those responsible for the planning and execution of sugarcane harvests; the burned cane must be reaped immediately, schedules readjusted, cane producers notified, and rolling stock and grinding machinery delayed, rerouted, or set in motion.[6]

Perpetrators of the unplanned or rogue fires that disrupt the smooth flow of canes to sugar grinding mills probably are motivated only slightly by a desire to destroy the crop itself. Such irresponsibility makes little sense in societies where sugarcane harvesting represents one of the few local sources of cash income.[7] Rather, the common interpretation for these fires is that they provide immediate employment, including important overtime wages, for the workers of a particular village, estate, or district. This interpretation, not incidentally, has been forwarded as the reason for unplanned Caribbean sugarcane fires for many decades. We must rely on these interpretations rather than testimony from the fire starters themselves because finding those responsible for setting illegal or unplanned fires always has been close to impossible; "detection of culprits is rarely possible, and even when successful, police and court action may fail because of unsatisfactory evidence."[8] This observation or lament—pointing out the impossibility of apprehending those responsible for present-day sugarcane fires—is, as one learns from reading early British West Indian newspapers or the reports prepared by sugarcane plantation managers in the early twentieth century, exactly the same as it was one century ago. The historical continuities in the motivations for starting Caribbean cane fires as well as the inability of authorities to catch incendiaries—along with the certainty that many of the sights and sounds and smells of cane fires today are the same as they have been for centuries—support the contention that a Caribbean sugarcane fire in the early twenty-first century opens a window on the region's past.

Sugarcane Cultivation in the Lesser Antilles in the 1890s

The British Caribbean sugar bounty depression of the late 1800s wrought more obvious changes on the landscapes of marginal cane-producing islands than it did in places where sugarcane continued to monopolize local economies. In January 1888 observers in Dominica (correctly) predicted that the island soon would cease to produce substantial amounts of sugar: "The area of most of our sugar estates is gradually being covered with cocoa and lime trees."[9] Five years later the widespread poverty and discontent there was blamed in part on the "many plantations . . . entirely abandoned."[10] The transformations under way on Dominica also had parallels in nearby islands, and not all of these changes led to distress; similarly mountainous Grenada had abandoned large-scale sugarcane production a few years earlier in favor of cacao farming. The relative prosperity among the small-scale cacao growers of Grenada, as far as the 1897 royal commissioners were concerned, suggested that a shift to crops other than sugarcane might be a model for the economic salvation of much of the region.

Yet dispensing advice and suggesting agricultural change was one thing; reorienting field alignments, drainage ditches, labor requirements, and work routines plus purchasing expensive and complex machinery and equipment were entirely another. In the British Caribbean in the late 1800s, a possible shift to nonsugarcane crops involved far more than simply discarding one kind of agricultural enterprise and infrastructure for another, because for nearly all of the islands, sugarcane was the only major cash crop they had ever produced. In an even broader sense, sugarcane production ultimately explained nearly all facets of the region's economic geography—including the nature and density of human populations, settlement patterns, road networks, and port facilities—because it was the activity that had created, beginning in the mid-seventeenth century, the character of each island in particular and the entire British Caribbean in general.[11]

Much has been made, appropriately, of the importance of the wealth accruing to European individuals, families, and nation-states that was generated by Caribbean sugar in the colonial era, economic success helping to explain why sugarcane came to dominate the entire Antillean region. Less well appreciated is that sugarcane's remarkable ecological adaptability created a veneer of economic and landscape homogeneity over a widespread and physically very diverse zone of Caribbean islands and rimlands.[12] To be sure, the subtropical latitudes of most of the small British islands provided the ideal climate for high sucrose content in sugarcane.[13] Yet beyond that generalization, the eastern Caribbean offered a wide range of environments in which sugarcane grew. The small drought-prone islands in the north, at about eighteen degrees north latitude, varied among themselves from flat and sandy Anguilla to St. Kitts with a volcanic mountain soaring 3,800 feet above sea level. Farther south, the volcanic Windwards exemplified startling interisland variety. Grenada's clayey soils, for example, were very different from the lighter and more porous soils of St. Vincent. Barbados, 100 miles to the east, was drought prone owing to its relative flatness, and Tobago, in part because of its small size, was a good deal drier than neighboring Trinidad. The mudflats of coastal British Guiana, on the shoulder of northeastern South America and only five degrees north of the equator, were, when compared with any of the islands, a different world altogether.

Within these overall regional differences, seasonal climatic changes also could pose a range of problems for Caribbean sugarcane growers. The hazard of too little rain—exemplified by the brutal drought in Barbados in 1894–95—could, within a few months, become that of too much, such as in June 1880 in St. Vincent, where dismally wet conditions turned the roads to mud and nearly halted the sugarcane harvest.[14] The annual hurricane season of late summer and fall produced unending anxiety, as it always had, for sugarcane planters. The huge hurricane of September 1898 flattened the growing sugarcane crops of Barbados and St. Vincent and dealt a glancing blow to St. Lucia's sugar industry on its way north. The storm also washed away the common wisdom among planters in the southern islands that hurricanes—because they had been so infrequent there for decades—now were confined to more northerly latitudes.[15] The following autumn the deadly "San Ciriaco" hurricane that ravaged Puerto Rico first visited the British Leewards, creating widespread devastation on St. Kitts and Nevis and bringing "many estates almost to the verge of ruin."[16] These severe late summer storms were far more serious, as they always are, during hard economic times. By the late 1800s, in part because of recent hurricane damage, not only was sugarcane finally on its way out in Dominica and Grenada, but it also recently had disappeared from Anguilla. The British Virgin Islands, Montserrat, and Nevis were not far behind.

In the islands where sugarcane continued to be the principal crop, local estates—influenced by technological changes and international sugar marketing transformations—were reorienting capital outlays, labor requirements, and production routines. Perhaps the most far-reaching modernization in the production of raw sugar in the eastern Caribbean was occurring in Trinidad. By the 1870s, heavy mechanical rollers were replacing manual techniques at some factories to hasten the throughput of canes. New centrifuge devices separated sugar crystals from molasses faster and cheaper than before. More and more, steam, not wind, powered Trinidadian cane milling, exemplified by the erection of the Usine St. Madeleine sugar factory directly east of San Fernando in 1872. The transportation of people, raw materials, and sugar products was facilitated by the construction of the new public railway network in western Trinidad that was augmented with a private, plantation-owned railway that shared gauge size, and therefore interconnectivity, with the public system.[17]

These innovations in Trinidad and elsewhere—such as the construction of central factories in St. Lucia patterned after those in the nearby French islands—occurred on lands where sugarcane already was growing. So the alterations were therefore incremental or piecemeal, unlike the sweeping landscape changes occurring at the same time in the Dominican Republic, for example, where American capital was transforming vast areas of scrub into cane lands. In any case, the resulting technological variety in British West Indian sugarcane production—ranging from the relic waterwheels in St. Vincent to the gleaming new vacuum pan equipment in British Guiana and Trinidad—among and within individual islands was roughly analogous to the region's environmental variation. The example of the several different types of sugar milling in tiny Tobago in 1897 highlights the wide variety in sugar production techniques there and, by extension, in the region as a whole. A table submitted to the visiting royal commissioners by the sugar planters in Tobago enumerated thirty-nine working sugar estate mills; twenty were powered by steam, eleven by waterwheels, five by wind, and three by the archaic system of cattle-powered grinding rollers.[18]

Ninety percent of the Tobagonian sugarcane acreage in 1897 was cultivated by "metayers" who tilled owners' lands as sharecroppers for a percentage of the cane's value. But this arrangement was not widespread elsewhere; rather, regional land and labor combinations, as in milling techniques and environmental characteristics, displayed wide variation. Traditional planter-laborer relationships and animosities persisted in St. Kitts, Antigua, and Barbados, where workers tilled the land for wages. As depression prices for raw sugar meandered lower in the 1890s, several planters on St. Vincent simply left their lands untilled (and the estate workers wageless) rather than adopt a sharecropping scheme. In nearby Trinidad the technical innovations of the late nineteenth century led to a notable consolidation of sugarcane land into ever larger estates. Yet the increase in Trinidad's overall milling capacity, combined with that island's relatively large size, also helped give rise in the early 1880s to a growing class of small-scale, independent cane farmers—both West Indian and East Indian—who grew canes on their own lands but who were nonetheless dependent on the large-scale sugar grinding mills nearby, where they sold their crops.[19]

Subtleties in methods of sugarcane field preparation also varied from one island to another. In the August 1902 *Empire Review*, Norman Lamont, one of the important estate owners of Trinidad, lamented that old-fashioned West Indian field operations had not kept pace with the introduction of new types of equipment. This lag in labor techniques, furthermore, needed correction (according to Lamont) if the British islands were to compete favorably with American sugarcane cultivated in the Greater Antilles. He asserted that even implements as fundamental as the plow, horse-hoe, and horse-drawn harrow were foreign to most estates of the region, a testimony in most cases to "the ultra-conservative ideas of the West Indian planters" and a situation that could be corrected only by efforts "to properly educate the rising generation."[20]

But a closer look at field operations elsewhere in the region would have supplied evidence that generalizations such as Lamont's were true only up to a point. To be sure, very old-fashioned agricultural techniques were typical on Barbados, where an underpaid oversupply of labor precluded the necessity for planters to invest in well-known innovations and thereby extended the traditional hand tillage of sugarcane into the twentieth century. But horse-drawn implements were in use in St. Kitts as early as the 1860s, and the introduction of the water buffalo to Trinidad had provided to that island an important draught animal found nowhere else in the region.[21] Even the mongoose, soon to be common to the cane lands in all of the islands for the eradication of rats and snakes, was in only a few places in the early 1880s. An account from St. Kitts in December 1883 announced, "We are glad to learn that the Mongoose has been introduced . . . a gentleman from England a few days ago having brought a couple dozen with him. The mongoose are very useful in protecting canefields from destruction by rats."[22] The writer's gladness about this biological introduction doubtless changed in succeeding years; a decade later, after the increased mongoose population had redirected its attention to local chickens and house pets, letters to St. Kitts and Antigua newspapers demanded its eradication.

The interisland differences in sugarcane production extended even to variations in local insurance schemes for the industry. In Antigua in 1894 fire insurance covered only plantation buildings, and only when they contained marketable produce. The sugarcane crop itself in Antigua traditionally was uninsured, being left "to providence." Barbadian planters, in contrast, were customarily less eager to trust their growing cane crops to fate, and an elaborate scheme of fire insurance existed there for estate buildings as well as for the growing sugarcane. Barbadian insurance premiums varied on a sliding scale according to whether an owner wished to insure a crop for the entire season, which involved paying higher premiums, or only the harvest period.[23]

Those who deplored the moribund old-fashionedness of British West Indian sugarcane cultivation were ignoring not only the recent technical innovations in Trinidad and elsewhere but also the contemporary accomplishments of agricultural scientists in the region. C. A. Barber, the superintendent of agriculture in the Leeward Islands, conducted and reported on a number of locally useful studies in cane disease and pest infestation on experimental plots in St. Kitts. The larger colonies such as Barbados and British Guiana had sugar "chemists," agricultural officers paid by public funds to conduct field experiments and advise about recent techniques. These men also occasionally traveled to nearby smaller islands for consultations about specific events such as pest or disease infestations. Barbadian agronomists J. R. Bovell and J. B. Harrison were regionally well known for their cane-breeding experiments; from existing varieties on Barbados, they developed disease-resistant strains that prospered locally and were subsequently successful elsewhere in the region.[24] The importance of these innovations in applied biology was, by the turn of the century, by no means restricted to crop scientists and government officials. When Daniel Morris, by now the head of the West Indian Department of Agriculture, lectured to the Barbados General Agricultural Society about the natural history of sugarcane in July 1901, he discussed many of the findings of Bovell and others. Morris's lecture was enthusiastically received by the governor as well as by a number of practically minded cane planters who only a few years earlier—prior to the proven success of Bovell's work—had considered such lectures the utterances of "so many parvenues issuing from academic cloisters and experimental laboratories to teach experienced men how to till the soil."[25]

Agricultural innovations in the British West Indies in the late Victorian era were influenced by as well as being an integral part of the rapidly expanding field of agricultural science worldwide. By the late 1800s, scientific botany overlapped with analytical chemistry, and applied botanical research routinely involved the use of laboratory equipment such as the microscope. A prototypical study by William C. Stubbs—the director of Louisiana's sugar experiment station near New Orleans—titled *Sugar Cane* appeared in 1897. Stubbs's book (which included a brief chapter describing Barbados's anachronistic system of hand tillage) provided detailed, microscope-enhanced diagrams and drawings that delineated and clarified some of the fine points of sugarcane physiology.

Among other topics, Stubbs discussed and emphasized the paramount importance of water in the growth and value of sugarcane as a cash crop, and he cited comparative studies from Martinique and Louisiana revealing that water constituted 70 percent of sugarcane's weight. Stubbs further pointed out that the amount and quality of liquid inside the stalk—which was, after all, the physical basis of sugarcane's economic value—varied crucially as to seasonal and even daily weather characteristics, owing to the plant's remarkable ability to take up and transpire water. But water's role in the growth and seasonal cycle of sugarcane was complex; it was not simply a case of always providing as much water as possible. Planters usually hoped for a dry harvest season because the reduction of excess water in the cane stalks then yielded a "sweet juice." Rainy conditions at reaping time, in contrast, would cause the cane to take up too much water, which diluted the sucrose content: "A good rain will make a difference in the juice of cane cut the day before and the day after the rain."[26]

Stubbs's work and subsequent research by modern tropical agronomists highlight water's importance in understanding sugarcane's growth. Their studies also throw light on the crop's remarkable environmental range, leading one to suspect that the plant's ability to cycle copious amounts of water through its roots, stalks, and leaves may provide a key to understanding how some of the small islands of the eastern Caribbean have sustained sugar crops for three and one-half centuries in a variety of terrains and soils. A. C. Barnes, writing almost one century after Stubbs, estimated that over a growing period of twelve to fourteen months, sugarcane requires three meters of water, an astounding quantity far higher, of course, than the final water content of the plants at harvesttime. With adequate water, sugarcane, according to Barnes and others, has "the ability . . . to adapt itself to a wide range of environmental conditions."[27] Nor is water's vital role in sugarcane growth unrelated to the crop's considerable labor requirements. Fields must be adequately prepared so that mature cane plants are afforded the maximum availability of rainfall during growth and then drained when excess water is undesirable. The arduous hand tillage on Barbados sugarcane estates in the 1890s, for example, involved covering fields with cane trash and leaves as well as digging two-foot-square holes for each cane plant to collect and conserve moisture. At the same time, however, drainage ditches had to be dug, and even tiled, for every four rows of planted canes to prevent flooding in case too much rain fell.[28]

The paramount (although imperfectly understood) role of water in the crop's growth is of course important in any discussion of either sugarcane's combustibility or the associated frequency of sugarcane fires. The high percentage of water in a stalk of sugarcane almost certainly provides a modest degree of fireproofing for growing crops and, in a broader historical sense, an ecologically oriented explanation as to why sugarcane has been such a durable, lasting, and dominant feature of Caribbean landscapes for centuries. Given adequate moisture, the volume of organic material produced by sugarcane is remarkable. According to G. B. Hagelberg, "Sugar cane offers an inexhaustible source of energy and a quickly reproducible raw material. . . . A sugar cane field is something like an annual forest in the volume of organic material it generates."[29] When one considers that this "annual forest," cultivated in the islands year after year for centuries, is composed of a plant that is, strictly speaking, a giant grass, one cannot help but wonder why it has not always fueled endless islandwide cane fires throughout the Antilles. Any adequate answer to this question, furthermore, would have to take into account water's importance in the plant's physiology and growth.

But partial physiological protection from fire provided no guarantee that, under the wrong natural or social conditions, sugarcane fields would not burn in their entirety, leaving planters poorer and workers even worse off. Island planters at the turn of the century took recommended preventive antifire measures, lest their planting techniques invite disaster. West Indian agricultural officials were well aware, for example, of the potential dangers posed by the crop homogeneity that sugarcane always represented. Writing in 1899, agricultural officer William Fawcett of Jamaica warned, in the face of growing threats posed by diseases, insect pests, and incendiarism, that when "plants of one kind are grown together on a large scale," such a technique could lead to "utter ruin to planters."[30] Five years later in Antigua, the virtues of the new "Zayas" cane-planting technique from Louisiana were being discussed actively among local planters. The main feature of the process was a wider spacing between individual canes, which allowed for better air circulation within the fields and assured that "the cutting of the canes will be easier and cheaper and cane fires will be less frequent and injurious."[31]

Using Fire in Sugarcane Cultivation

Fawcett certainly was no pyrophobe, because he was the official who, two years later, would remind a gathering of regional agricultural officers in Barbados that seasonal burning by local shepherds in the islands often improved rural grazing pastures and livestock foraging. On occasion, but rarely without reservation, British Caribbean sugarcane planters of the late 1800s also used fire as part of their agricultural routines.

Observations and commentary both for and against using fire in sugarcane growing came to Kew Gardens in written reports from the West Indies and elsewhere, advice and opinions that often were conflicting and which varied as to terrain, scale of operation, and latitude. By 1897 in Louisiana, for example, cane fields were routinely burned after the autumn harvest; this was done to eliminate the accumulated trash that would otherwise become soggy during the winter rain and thus inhibit the regrowth of cane stubble in the spring. This strategy took into account a loss in potential nitrogen from burning the cane leaves.[32] In the British Caribbean, using fire in cane cultivation—a technique on which planters may have relied more and more from observing local cane farmers—was becoming more widely accepted after the turn of the century. In 1911 Noel Deerr, while noting the loss of nitrogen when cane trash was burned, highlighted the benefits of burning after harvest in the West Indies and elsewhere, a technique that he acknowledged was barely tolerable "to agriculturalists accustomed only to European farming practice."[33]

Of course the burning of diseased and pest-infested crops, including sugarcane, was, as in forest clearance, the only expedient way to eliminate unwanted organic refuse. As the growing canes in the West Indies came under increasing attack from fungus-borne diseases at the end of the nineteenth century, there was little recourse except to burn the affected plants. A fungus "rot" first identified in Barbados in 1890 spread to most of the island within three years and then to St. Vincent by 1894; in both places the authorities adopted widespread burning practices to counteract it, but with great reluctance.[34] In 1896 in St. Kitts, an estimated 8 percent of the island's sugarcane crop was ruined by disease. The local planters hoped that importing the Caledonian Queen and White Transparent varieties from Jamaica would provide some inherent resistance for future crops. But the affected Bourbon canes on the island were burned.[35] As sugarcane gave way to other cash crops elsewhere in the region, burning was often recommended by local agricultural officials to eliminate diseased plants, as in the case of the Sea Island cotton planted in the Choiseul district in St. Lucia in July 1907.[36]

As was widely practiced among rural subsistence farmers in the region, burning was usually the first—and in some cases the only—line of defense against insects on sugarcane estates. Newspapers and agricultural reports commonly reported the firing of cane fields to reduce insect infestation, if for no other reason than that it usually sent the pests flying into neighboring cane estates. Professor J. A. d'Albuquerque, Barbados's sugar chemist, identified a new and harmful beetle attacking local canes early in 1892. Unlike other types of cane borer, this insect passed its entire life cycle inside the plant and therefore had to be eliminated by burning: "We shall advise that the standing crop be reaped as expeditiously as possible, and that the trash and tops be burnt in the fields immediately. . . . No chemical preparation can be more efficacious or so universally applied as fire. The ancients knew well the value of fire, and burnt the stubble on their fields for the double purpose of destroying noxious agents, and of liberating potash, thus purifying and manuring at one and the same time."[37]

Few other commentators were so unequivocally positive about firing sugarcane to control insects and diseases. The written opinions early in the 1890s nearly always agonized over the prospect that using fire might create even worse problems, mainly because the cane trash—which inhibited weed growth and supplied texture to the soil—would be eliminated. When a particularly serious outbreak of shot borer threatened the St. Kitts sugarcane industry in 1892, C. A. Barber, the superintendent of agriculture of the Leeward Islands, told planters to heap up the rotten canes and burn them, but "do not burn the trash, as you will therefore destroy your friends the ants."[38] Similarly, the (reluctant) decision to use fire to combat the fungus disease in the St. Vincent cane fields in 1894 came only after a meeting of key planters in Kingstown who agreed that "dire necessity" called for immediate yet—because fire was involved—risky action. The Vincentian planters feared that excessive burning of cane trash there would reduce the quality of the island's porous volcanic soils.[39] Even when agricultural experts called for burning, locals were not always eager to carry it out. In early 1895 in Antigua, a combination of disease and pest infestation threatened the growing sugarcane crop, but the island's planters wondered if external advisors took their particular situation into account: "Our planters have been charged by Kew with not taking intelligent pains to combat the enemies of sugar cane; but we have adopted all the recommendations . . . except the wholesale application of fire to the fields—a proceeding which our natural conditions would render suicidal."[40]

But in the next several years, antiburning orthodoxies among the sugarcane estate planters of the British Caribbean were relaxed, even to include the adoption of firing canes prior to harvest. In British Guiana in 1897, according to government chemist Professor J. B. Harrison, a key to understanding the relative success of the local cane outputs was that field workers invariably plowed (unburned) trash composed of dead leaves and cane tops back into the soil after harvest.[41] But that changed during the next decade or so; by 1911 the widespread practice along the Guianese coastlands called for the sugarcane "fields [to be] fired immediately before harvest" because the preharvest burning reduced expenses. Planters saved an estimated five to six shillings per acre in canecutters' wages, as well as the labor cost required to bury the cut cane leaves after the crop was finished, if the leaves were burned off before harvest. Burning the cane trash in British Guiana now was interpreted as having beneficial ecological side effects; the practice eliminated "fungus spores and noxious insects" and, by allowing the ash constituents of the cane trash to fertilize the soil, encouraged the subsequent "ratoon" crops of young canes to come up quicker.[42]

Attitudes about cane burning were changing in the islands too. "Authorities" were setting preharvest sugarcane fires in Trinidad as early as May 1899, a strategy that possibly came from conventional agricultural-scientific exchanges with British officials elsewhere or perhaps from observing the success of local Indian rice planters who always fired the stubble remaining from the previous year's crop in their fields before planting.[43] In 1902 in Basseterre, St. Kitts, Francis Watts, in his lecture to a group of agricultural teachers, alluded to the "regrettable practice of burning" that, in his estimation, reduced the soil quality, and he warned his students "never to resort to such a practice except for the destruction of . . . weeds which might not be got rid of in any other way."[44] Watts likely was countering a growing sentiment in the region that on sugarcane plantations, "fire is beneficial to a field and that good crops are grown in the years immediately following . . . fire," an idea gaining ground in Barbados by 1906.[45]

Changing attitudes toward preharvest sugarcane burning in the British Caribbean islands reflected not only local ecological issues but also the impact of wider socioeconomic trends and changes. The ongoing competition from European beet sugar and from the new, large-scale sugarcane operations elsewhere in the tropics meant that any way to reduce costs—such as burning sharp cane leaves and thereby reducing canecutters' wages—was gaining more acceptance than before. The increasing dependence on imported organic and inorganic fertilizers on the largest estates in the region lowered the concern over nitrogen loss through burning cane trash. Perhaps most important, job possibilities elsewhere were opening up for insular labor forces. By 1906 sugarcane planters throughout the islands were bemoaning the loss of their best field workers to the Panama Canal and other circum-Caribbean projects—such as canecutting in Cuba and banana work in Central America—financed by North American capital. So in the British islands, preharvest sugarcane burning, along with other laborsaving techniques, likely came to be considered a necessary, if distasteful, step to bring in cane crops with reduced labor forces, and it perhaps represented a way to reduce the arduous character of canecutting so that young men might decide to stay at home.

Whatever the reason, preharvest burning had become common in the British Caribbean islands by the 1920s, especially for some of the newer cane varieties.[46] In subsequent decades the intentional firing of ripe canes prior to reaping has become common practice in most sugarcane-growing regions of the world. The improved field condition that cane burning brings for manual canecutters is now coupled with heightened concerns about the subsequent weather conditions, because burned canes deteriorate far more quickly after a rain than do unburned stalks.[47]

Whereas the subject of sugarcane burning was open to debate, no plantation operators ever questioned the need for sugar factory fires. Fire was the catalyst for the production of cane sugar from earliest times. The excess liquid from the canes had to be evaporated, leaving the semiliquid sucrose as the elementary ingredient for either molasses or unrefined sugar. By the end of the nineteenth century, since firewood was so scarce throughout the region, its potential use as boiler fuel was very limited. Further, the gathering of local wood involved competition with peasantries who gleaned sticks and occasional logs for their cooking fires from forested interiors. Yet in rare instances, such as in Antigua in October 1889, estate laborers continued to collect brush and firewood to help fuel sugar boilers.[48]

The Antiguan laborers were gathering the natural firewood to compensate for the rare shortfall in local sugarcane "megass" because of drought in the preceding year. "Megass"—better known as "bagasse" and also spelled "megasse" or "megas"—was the general term for the fibrous material in the pressed cane stalks that remained after the extraction of juice at the mill. In an earlier era, bagasse fueled the so-called open pan system of juice boiling that followed after wind-powered rollers (still very common on Barbados and some other places at the turn of the century) crushed the juice from the canes. After the advent of steam power, "the bagasse was used as fuel for the boilers, a practice which continues extensively under modern conditions."[49] By the first decade of the twentieth century, bagasse represented the only industrial fuel in West Indian sugar factories, and its combustion fired sophisticated "multitubular boilers" as well as giant iron furnaces built for the special purpose of burning "green bagasse."[50]

The indigenous fuel supply of crushed cane fibers for the Caribbean sugar industry often provided too much, as opposed to not enough, combustible material. Despite the rare seasonal shortages (such as in Antigua in 1889) the accumulated bagasse—plus the harvested tops and leaves—more often presented disposal problems for sugar estate managers. As bagasse surpluses accumulated, factories in some of the islands sought solutions by burning "a greater quantity than actually required," an industrial waste disposal strategy that eliminated the surplus fibrous material but at the same time sacrificed heat efficiency in the factory boilers. Sugar manufacturers also used surplus bagasse for rum distillation.[51] Of course, the bagasse not immediately used as fuel or plowed back into the fields had to be stored someplace. So after the sugarcane was crushed at the mill, the bagasse by-

The interior of a sugar boiling house in Barbados about 1900 (from Ann Watson Yates, *Bygone Barbados* [Barbados: Blackbird Studios, 1998])

product usually was stacked, dried, and then stored. Megass or bagasse buildings and sheds, where the dried fuel was kept, were ubiquitous features at the settlement nucleus of every Caribbean sugarcane plantation. Traditionally the bagasse houses or sheds were wooden frame buildings, but by the turn of the century they were often constructed of corrugated iron.

The West Indian newspapers of the time told of one accidental bagasse house fire after another, an unsurprising (and almost certainly underreported) observation when one considers that these structures sheltered stockpiles of dried and highly flammable material. Laws on every island were designed to prevent accidental bagasse fires. These statutes specified severe penalties for persons carrying open flames or lit pipes into bagasse storage houses, and they were also legal testimonies as to the frequency of ruinous fires. A typical bagasse fire at Buckley's Estate near Basseterre, St. Kitts, in June 1884 was brought under control "by the prompt assistance rendered by the laborers" on the estate with the help of a police detachment from the nearby capital town.[52] Nor did bagasse have to be under shelter to represent a fire hazard. During the milling season, piles of the material were common sights around plantation yards and estate factories and were susceptible to ignition from any spark or flame. On March 10, 1882, at Warren's Estate in St. Michael parish, Barbados, eleven heaps of dried "megas" burst into flame from an unknown source during the midafternoon and burned far into the night.[53]

The occasionally destructive accidental sugar factory fires in the countrysides—some of which must have been truly spectacular events—sometimes occurred because fire-prone bagasse houses were too close or even attached to the larger buildings. But usually it was impossible to pinpoint the origins of these fires. At midnight on March 29, 1892, at Belle Plaine Estate in St. Andrew parish, Barbados, the dry material on the roof above the estate boiler room caught fire. The only water available was being used to clean the evaporating pans, and observers marveled that the damage was not more extensive.[54] Haynes Hendy, the proprietor of Burleigh Castle Estate near Scarborough, Tobago, was far less fortunate the following year. On the evening of September 4, 1893, Burleigh Castle's newly renovated plantation factory (the building and equipment insured for £1,000) burned to the ground. The blaze was detected about 9:00 P.M., and soon the entire building was enveloped by flames. Any attempt to save the factory was abandoned: "At ten o'clock, about which time the weighty roof fell in with a tremendous crush, a stream of fire shot upwards, the night glare from which . . . gave the appearance of a bright moonlight night. A few minutes thereafter nothing remained of the once-extensive building but a mass of smouldering embers."[55]

Sugar factory fires and bagasse house fires, because they occurred in rural areas, normally attracted fewer onlookers than did urban fires. But when large estate fires were close enough to towns and cities, crowds naturally appeared. When the Wall House plantation, about a mile outside Roseau, Dominica, caught fire late on a Saturday night in mid-January 1886, hundreds of people, most on foot but even some of Roseau's white elite on horseback, followed the shouts and flames to the conflagration. Yet despite the size of the gathering, they were too late to save the building. They "could scarcely do more than look at the grand, but terrible scene of the entire building, full of last year's Megass, in one fearful blaze."[56]

"Night after Night": Incendiarism in the Cane Fields

The newspaper description of the burning of the Wall House plantation on Dominica concluded, "There is not the slightest doubt that the fire was wilfully set." While it cannot be proven (any more than the 1886 newspaper assertion about the Wall House fire could be proven), it is likely that well over half of the bagasse house fires in the fin-de-siècle eastern Caribbean were the results of arson. These fuel shelters were natural targets for "a certain class of laborers [who] have some grievances real or fancied to redress," as in St. Vincent in July 1881, where several bagasse houses were torched. Although estates and insurance companies posted notices for information leading to the arrest of those responsible for the St. Vincent fires, typically "no one comes forward to claim the reward."[57]

Reports of bagasse house arson abounded in newspapers and official gazettes. Accumulations of the dried cane residue posed tempting targets for incendiaries, whether or not the material was under cover. In June 1885 in St. Kitts, "megass heaps" were set ablaze by "malicious persons" whose ire apparently was directed toward an unpopular estate owner.[58] Indeed, with so much flammable material lying in heaps around plantation yards during the sugarcane harvest seasons, the temptation to burn may have bordered on irresistible. Two seven-year-old boys were accused of firing cane trash at Bruce Vale Estate and taken before the local magistrate in May 1882 in St. Andrew parish, Barbados; they were subsequently discharged because they were so young.[59]

But the intentional burning of bagasse houses and the burning down of the factories themselves were, in more ways than one, very different acts. The plantation factories, which housed milling and boiling equipment, were more difficult to ignite and, unlike bagasse sheds, would have been occupied by workers and supervisors—at least during harvest—around the clock. Less obvious is that the torching of plantation factories, while providing the momentary satisfaction that a despised estate owner or manager would lose property and profits, shut down milling and grinding and thereby deprived estate workers of wages. Nonetheless, even though bagasse houses were far more often the objects of incendiaries' actions, in the explosive atmosphere of the late 1800s even factories were maliciously burned on occasion. In late June 1884 the entire factory along with its machinery on Langford's Estate in Antigua was lost to arson, and observers commented afterward on the level of collusion necessary in such a small community, where everyone knew everyone else, for such a heinous crime to produce no witnesses.[60] Ten years later, also on Antigua, the entire works of the Union Estate, along with barrels of raw sugar and puncheons of molasses, burned, almost certainly "the work of an incendiary."[61]

Another way in which spiteful workers could secretly settle scores with hated owners and planters without putting local jobs in jeopardy was to injure or kill their livestock. These crimes, perhaps more often than not, were coupled with fire. Grenada in the 1890s provided three such examples. Early in November 1892, unknown "ruffians" cut the throats of four mules and disemboweled the animals at the Mount Craven Estate in St. Patrick "and then set fire to the megass house."[62] The poisoning of a pig that was owned by Simpson Marucheau was coupled with the burning of Marucheau's cocoa shed at Snug Corner three years later.[63] In July 1897 the "spirit of incendiarism" in the air throughout the St. Andrew's district of Grenada was said to intoxicate the laborers, inspiring them to burn at random and, in one case, seriously injure a prize ox at Mirabeau.[64] One month after the fire-accented 1896 riots in Basseterre, St. Kitts, someone set the cattle shed ablaze on an estate near Dieppe Bay, at the northern end of the island, after midnight: "There were twenty-five cattle in it. . . . One managed to burst the chain and . . . the other twenty-four were burnt to death. Some pitch pine lumber stowed away in the shed was destroyed also. The bellowing and struggles of the poor animals, and the screams of the women present . . . created . . . a scene of indescribable horror."[65]

Incendiaries, however, targeted animals and buildings far less often than they did the cane fields themselves. Cane fires occurred on every island; a few were accidental, but probably most were set purposely. All the inhabitants of the eastern Caribbean in 1900, moreover, were as familiar with the purposely set or "malicious" sugarcane fire as they were with their own places in the region's socioeconomic hierarchy, subjects that were by no means unconnected. Blazing canes, just as they represented group identity, slave origins, and assertiveness in the fiery *cannes brulées* processions in towns and cities, symbolized a good deal more than crop combustion in Caribbean countrysides. Cane fire frequencies indicated—indeed marked and underscored—levels of social satisfaction and economic well-being among an island's working peoples. Contented laborers and their families with enough food to eat and who received reasonable estate wages set few cane fires; those less well off often burned fields. Those fires of despair were then replicated nearby with others set for the same reasons, perhaps not that day or that night but soon thereafter.

Everyone must have understood that the cane fires signified desperation. Even though the sugar bounty depression created general hardship throughout the region for twenty years, some seasons in some years were worse than others, conditions ultimately based on local circumstances. Authorities in Antigua in March 1895 acknowledged that the low-lying island was passing through "a crisis of almost unprecedented severity," what with European sugar bounties, drought, crop shortfalls, and then "several fires in different parts of the island," which there was far "less difficulty in determining the origin of than to discover the perpetrators."[66] Two years later in Barbados, cane fires pinpointed a more localized yet no less desperate scene. A potato raid or "riot" at Boscobel In April 1895 led to arrests of hungry rural dwellers who had taken plantation crops. The "incendiarism [was] said to be reprisal for the convictions and punishment of the rioters."[67]

Although it must have been transparently difficult, island officials seem to have tried their best not to acknowledge, at least publicly, that the cane fires—which at times literally burned all around them—signified widespread and volatile discontent. This brazen official disregard for the obvious was exemplified in 1901 in Barbados. From every account the sugarcane fires were particularly numerous and damaging that year; newspaper reports and official correspondence were filled with descriptions of, among other topics, the frustrating "conspiracy of silence" among rural tenantry dwellers who refused to identify incendiaries. In one memorandum to London, local officials spelled out clearly why they did not conduct formal investigations into the reasons for and causes of the fires, a provision that had been stipulated by the Fire Enquiry Act of 1879: "It was considered undesirable to enquire into them too closely, it being feared that persistent enquiry would lead the labouring population to think that they had only to set fire to a cane field to ensure their having an opportunity afforded them of making known their real or supposed grievances."[68]

Not every sugarcane fire was purposely set. Railway sparks, occasional flashes of lightning, and carelessly discarded embers or cigarettes could and did ignite accidental cane fires. So local prohibitions were designed to reduce open flames in or near ripened cane fields. So-called day sleepers in St. Lucia who illegally went "torching for crabs" at night set a cane field ablaze in July 1885, causing "sad losses to many daily hard working people."[69] The following year in St. Kitts, Robert Michael was found on Shadwell plantation with matches, cooking utensils, and food on a small patch of ground that he had cleared in the middle of a cane field; Michael was taken before the local magistrate and charged with "having had a fire in a standing piece of cane."[70] Both the St. Kitts and St. Lucia cases provide examples of how the enforcement of local statutes protected growing canes from accidental fires, and the narrative accompanying each report demonstrated that, at least in officials' minds, inferior people and the unsupervised use of fire were twin evils that often accompanied each other.

Cane fires were sometimes dismissed officially as pranks, usually when they were set by children, such as in Barbados in 1901: "Several fires have been caused out of sheer mischief, and there are two or three lads now in the Government Reformatory who are there for having set fire to Cane Fields."[71] The commonly used "mischief" caricature allowed officials to avoid the obvious point that these "lads" probably were indulging in behavior they learned from adults. One did not have to go far from characterizing those who burned cane fields as mischievous, childish, or irresponsible to describing the supposed excitable, intemperate nature of the black laborers themselves. Similar to the descriptions of fires in towns and cities, portrayals of flaming cane fields often depicted behavioral catalysts capable of unleashing and exposing primitive "African" emotions. The report of a sugarcane fire just outside Basseterre, St. Kitts, in February 1901 was typical of many such accounts; the flames attracted a "crowd of about as lawless a set of savages as can well be imagined. These people . . . were dancing and yelling in the most excited manner and howling imprecations on . . . the manager of the sugar plantation on which the conflagration was raging."[72]

Whereas dismissive, commonly practiced newspaper strategies publicly depicted cane fires as phenomena produced by primitive or childish people, very different terms, such as "criminal," "devilish," or "vicious firebrands," were more common in formal communiqués to London. While it is perhaps making too much of semantics, especially when meaning and connotation can change, these latter terms suggest that the local authorities knew exactly what they were up against when a series of sugarcane fires threatened an island's countryside. Disgruntled workers bent on revenge, not stupid or childish people, controlled the roads and pathways at night, their inflammatory activities reinforced by a code of silence among their fellow villagers. During the day these same nocturnal incendiaries melted into the ranks of plantation workforces. One newspaper description after another told the same story time and again. A mysterious fire, usually next to a road, would break out in a ripe cane field. When authorities arrived and questioned those in the vicinity who might have witnessed the event, no one could recall having seen anything suspicious. The next morning a harried estate manager hastily assembled canecutters to reap the burned canes before they soured. "Canefires, canefires, night after night and week after week is a burning disgrace to the labourers of this colony," moaned a tired-sounding Barbados newspaper writer in 1901.[73]

If the clandestine torching of sugarcane fields was attempted during the wet season (June to December for most of the region), very few such instances were reported. In the dry first half of the year, on the other hand, cane fires raged. A lit match thrown into dry cane trash usually could start a fire that would spread, but police inspectors occasionally reported finding kerosene-soaked rags where the fires had begun, especially in the early 1880s before matches were commonly used. After ignition, the crop's stage of growth and the prevailing atmospheric conditions often determined how quickly the flames would spread. Multiple attempts by incendiaries to ignite the cane fields at Whitehall Estate in Barbados early in January 1890 left the plants only scorched because the crop was young and green.[74] In March 1902 in St. Kitts, in contrast, a fire begun by an incendiary in the middle of the night at Pond Estate was started when "the wind was blowing half a gale." The fire "spread with amazing rapidity and a long stretch of field was soon ablaze." By morning eighty-seven acres of Pond's ripe sugarcane was burned. "The flames from these fields made a spectacle seldom equaled in a canefield fire."[75]

Yet the social rather than the physical atmosphere probably had more to do with the chain reactions that sometimes spread cane fires from one estate to another. How else could one explain the vividly described scene in Barbados in February 1889? "We regret to say that cane fires have been very numerous. Almost in every direction, night after night, the sky has been illuminated by the lurid glare from cane pieces."[76] It is also likely that fires of outrage and resistance to economic depression and oppressive social circumstances were in many ways regional, not simply local, phenomena. The gigantic waterfront fire in St. Lucia in August 1899 occurred when cane fires were raging in neighboring Barbados. Although St. Lucian authorities did not publicize the seemingly obvious connections between the cane fire and the wharf fire, they had little problem surmising the origin of the waterfront catastrophe and other fires in town: "Barbadian wharf scum . . . has been pouring into this place recently; and . . . they are being credited with the fires and robberies which have suddenly burst upon the accustomed quiet of this law abiding community."[77] Shortly thereafter in Castries, special constables were sworn in "as a measure of precaution against a repetition of incendiarism which had been openly threatened."[78]

Threatened or not, the sugarcane fire was not always a random outburst of frustrated despair. Probably more often it was a calculated act of vengeance or reprisal. Cane fires commonly appeared on estate lands after the owner or manager punished or disputed wages with field or factory workers. The resulting fires in the aftermath of such incidents seemed almost predictable. So it is hardly surprising that certain plantations—perhaps more accurately, the plantations' managers or owners—became the targets of incendiaries' wrath time and again. Pond Estate, near Basseterre, St. Kitts—the same place that would burn with windswept fury in 1902—was recognized as early as 1883 as a place that "for some years past . . . has been visited with a succession of fires."[79] Twenty years later "the frequent fires on Cane-fields on Gambles and Mackinnons" plantation lands in Antigua were interpreted as the vented outrage "of an incendiary."[80] Perhaps the most convincing evidence that fires were directed toward certain estates and their operators was that, amidst widespread fire epidemics, some estates seemed totally immune from the danger because they never were torched, as in Barbados in early 1891: "There are estates in all the parishes that are exempt from cane fires—their conductors must have discovered the way to secure indemnity from the evil—would not a comparison of notes on the subject help to check the growing evil?"[81]

In his letter to a Barbados newspaper early in 1900, C. J. Manning underscored the authorities' inability to apprehend cane field incendiaries, a problem common to every island in the region: "Night after night the cry of 'fire' rings in our ears. We are startled from our sleep, to see sometimes three, aye four fires raging at once. This . . . has been a record year for the number of these lawless acts . . . not one of which has been detected by the police."[82] Next year the Cane Fires (Prevention) Act of 1901 called for the punishment of convicted offenders as well as rewards for those turning in fire starters.[83] The bill had no immediate effect; cane fires continued unabated before, during, and after the legislative debates. No one except small boys were ever caught, but if incendiaries could be apprehended, punishment would not be far behind. Spirited exchanges in Barbados weighed, for instance, the merits and demerits of flogging convicted incendiaries. One supporter of flogging seriously doubted that the use of the "cat" would usher in, as some suggested, a new era of barbarism and darkness, because even such darkness would be preferable to "the light proceeding from burning canefields or buildings!"[84] But what difference did the severity of punishment make if offenders could not be convicted and then punished? One wag offered that fire prevention legislation calling for perpetrators to be "decapitated" still would be useless, for all the good it would do in stopping sugarcane fires.[85]

Barbadian authorities tried many approaches, nearly all in vain, in their attempts to apprehend sugarcane incendiaries. They printed circulars, posted placards, and offered rewards. The police department rescheduled constables' assignments and sent detectives—whose whereabouts probably were well known to black tenantry dwellers—to watch cane fields at night. Why could no one ever be found to identify the criminals? The answer, of course, had much to do with antiwhite solidarity among the closed-mouthed black Barbadians.[86] In any case, when standard policing procedures failed among an unresponsive and hostile citizenry, officials decided that the cane fire epidemic seemed to be, after all, a moral and ethical issue and pleaded with clergy to denounce it in their sermons. At several points, the island's authorities attempted (also with little or no effect) to push the antifire issue through "education." Rural teachers were urged to inform schoolchildren of the evils of cane burning. Perhaps, went one point of view, curricular change in schools could curb "the recurrence . . . of the cruel canefires" by redirecting Barbadian schoolchildren away from "useless knowledge" and toward the agricultural education of "heart, and head and hand as Professor Booker T. Washington is so practically doing at his schools at the Tuskegee Institute."[87]

Yet West Indian workers who started illegal cane fires were neither ignorant of the practical implications of what they were doing nor always motivated by vengeance. The characterization of all sugarcane fires as "malicious" and "evil" obscured the point that, since the crop required cutting immediately after burning, the fires usually provided instant harvesting jobs for poor black workers. Planters in both Antigua and Barbados early in 1895, for example, acknowledged that cane fires probably were set—at least in part—to "force grinding operations," and they commented on the lamentable reduction of sucrose when unripe canes were burned and then reaped.[88] In some cases, such as when these events took place in peri-urban areas, incendiary fires gave women and children the opportunity literally to invade the fields and to carry away the canes for their own uses.[89]

The ploy of setting fires to produce harvesting jobs was probably more common than the existing archival records would lead us to believe. It was far easier for officials to issue histrionic reports and commentary about "incendiaries" than to acknowledge the lengths to which desperate people would go to get work, both regular harvesting jobs as well as overtime wages. Contemporary suggestions about changing the way cane crops were insured on Barbados bear this idea out. Although most of the island's planters insured their sugarcane crops on a sliding scale, the insurance premiums—whenever they were paid—usually did not cover the value of the entire crop. One fire prevention suggestion in 1900 recommended that the planters pay higher premiums so that insurance companies could pay the "full value" for a burned crop, which would then be "allowed to stay on the ground and rot" rather than having canecutters reap the fired canes. "By doing this the fire fiend would soon find out that instead of getting cane to suck and liquor to drink, he would be putting himself out of work."[90]

Of course the insurance companies would have had to agree and even take the lead for such a strategy to be adopted, yet they were understandably unwilling to do so. When the cane fires act was debated in Barbados early in 1901, one suggestion called for burned canes not to be harvested but left in the ground, an idea that backfired. A deputation of local representatives of insurance companies met with Governor Frederick Mitchell Hodgson, the island's attorney general, and the chief of police. Not only were the insurance agents not eager to join planters by changing their conventional business arrangements to help stop the fires, but they also indicated that if some successful antifire action was not taken soon, "the Insurance Companies would have to consider their position and decide whether Insurances could be continued on existing terms or continued at all."[91]

Whereas the overall number of rural cane fires appears to have been much smaller in Trinidad than in Barbados at the turn of the century, a handful of incendiaries in the former place, unlike in Barbados, were identified and apprehended. The reason, moreover, seems to have been similar to the explanation as to why urban incendiaries were caught more frequently in Port of Spain than in Bridgetown: Individual members of competing, working-class ethnic groups, showing little intercultural solidarity, apparently turned in arsonists from other groups as a matter of routine. Late in 1897, for example, four black men were apprehended for torching a Chinese shop in the Caroni district of central Trinidad, an incident similar to those occurring at the time in Port of Spain.[92] The description that follows, furthermore, provides rich detail as to how a black incendiary was caught by Indian watchmen in a Trinidad cane field four years earlier. This single and perhaps isolated instance cannot explain completely the differences in either interisland cane fire frequencies or the apprehension of incendiaries on the basis of differing ethnic patterns. Yet it shows the very real social differences in the rural areas of Barbados and Trinidad at the turn of the century.

On March 7, 1893, Charles Ifill (whose name suggests that he was probably a Barbadian immigrant) was arraigned in Princes Town for firing a cane field at the nearby Cedar Hill Estate; he was eventually imprisoned for the crime. Cedar Hill, a property owned by the Colonial Company, had seen a number of fires early in 1893, so the manager hired watchmen to try to prevent further acts of suspected arson. On February 22, two sentinels, Abdool and Sankarsing, followed Ifill as he crossed the plantation boundary and headed deep into the sugarcane fields. Then they hid and watched him heap up dry cane trash and light the pile with a match. Ifill paused to make sure the blaze was spreading and then turned to leave. At that moment the Indian guards grabbed him and took his cutlass. They were holding him when the plantation manager arrived. The watchmen knew Ifill as a former worker on the estate, and at the court proceedings they identified him by his "swollen hand and crooked legs and feet." Three other Cedar Hill laborers—Pearah, Jokoo, and Teeluck—corroborated the testimony and "identified prisoner whom they swore to having seen on the spot where the fire occurred."[93]

Within the overall context of the sugar bounty depression, the epidemic of sugarcane fires in the Lesser Antilles as the twentieth century opened may well have been the British Caribbean's most serious single problem. In Barbados, the most severe case, the fire emergency preoccupied members of the government as they sought to legislate the fires out of existence, and the fires pushed Barbadian newspaper editors and letter writers to verbal extremes in their attempts to proclaim the dire circumstances they faced. More important than newspaper bombast, the West India Committee in London in March 1901 expressed direct concern about the Barbados cane fires to the British colonial secretary, Joseph Chamberlain. They urged him to intervene directly by contacting the island's governor to take preventive action, expressing the fear that "should these fires continue unchecked, the Insurance Companies will refuse to take risks over the crops and . . . the security of those mortgages."[94]

The regional importance of sugarcane fires could not have been stated more bluntly than by the letter Malcolm Kearton & Company in London sent to all of the major newspapers in the British Caribbean in May 1898: "The continued recurrence of incendiary fires in the West India Islands is a matter of grave concern to those who, like ourselves, have a large amount of capital invested in West Indian properties." Citing the recent unsolved fire at Alma Estate in Tobago as only one of a number of such cases, the letter continued to warn that the London "Insurance Companies, we believe, will see fit to charge almost prohibitive rates unless there is a discontinuance of these incendiary fires." This grim warning, which threatened the capital investment climate for the entire region and the security of property already in place, could not have been more ominous. While one Trinidad newspaper considered the warning overblown, pointing out that relatively few fires had recently occurred there, the Kearton letter must have shaken many planters and officials, reinforcing their resolve to fight the fires.[95]

Soon thereafter the incidence of cane fires declined in Barbados; according to annual police reports, the total number of cane fires in 1902 (74, an estimated 45 of which were set by incendiaries) dropped dramatically from the preceding year (236).[96] Since the vast majority of sugarcane fires always occurred in the relatively dry first half of any calendar year, the drop in fires reported in 1902 was obvious by April. The initial interpretation was that the recent cane fire prevention law, which called for substantial monetary rewards for information, apparently had done some good.[97] But no surge in either the number of informants or arrests of incendiaries corroborated this interpretation.

Why did the number of Barbadian cane fires decrease in 1902? The Brussels Convention in March of that year rescinded the European beet sugar bounty system, but the trickle-down effect in terms of better sugar prices leading to better workers' wages was not felt for some time. Perhaps Barbadian cane fires declined because the slight reduction in import tariffs in 1901 had put more and cheaper food into the stomachs of local laborers. Maybe the island's smallpox epidemic of 1902–3—which caused 119 deaths among nearly 1,500 cases—played a role. One benefit of the smallpox epidemic, as far as black workers were concerned, was that it prevented the export from Barbados of "large quantities of sweet potatoes and other produce" to nearby islands, food that "therefore entered into local consumption."[98] Although unlikely, perhaps authorities and newspaper publishers agreed not to report sugarcane fires, hoping to downplay the menace so that they themselves would not be spreading the fires by publicizing them. Perhaps it was a combination of all of these and other factors.

Questions such as these about Caribbean sugarcane fires from any era perhaps never can be answered satisfactorily because of the one-sided nature of the evidence surrounding them. Planters and officials and newspaper editors, those who reacted to and described and interpreted the blazes, carried on conversations with one another about the fires and left their remarks for posterity. Those responsible for the fires, on the other hand, were almost certainly not silent about them at the time, for they doubtless had secret conversations among themselves, and one might argue that they let fire speak on their behalf. Yet for obvious reasons their actual conversations are lost forever. In any case, the sugarcane fires of the region, because of the social context in which they occurred, seem to provide an intriguing contrast between those who write history and those who create it.

It is also possible, even likely, that illicit cane burning at the turn of the century played a formative role in shaping the region's future conventional cane harvesting technology. In the 1890s, on some of the islands and on some of the estates where incendiary cane fires seem to have occurred over and over, plantation managers and laborers must have become used to reaping burned canes, noting, perhaps, some of the benefits of burning.

Shortly thereafter, although preharvest cane burning seems to have become common in the region only in the relatively large colonies of Trinidad and British Guiana, the idea of using—rather than condemning—fire in sugarcane cultivation was gaining ground in the region. At the very least, it is an appealing notion ecologically to think that burning—the ultimate act of worker resistance against the plantation crop that created the colonial Caribbean—has today become embedded in the Caribbean region's sugarcane cultivation routine.

6

Fire and Water

NO ONE COULD BLAME CAPTAIN WALTER S. Darwent for the water shortfall during the huge fire that destroyed so much of the commercial property in central Port of Spain on March 4, 1895. Darwent, the chief of the city's fire brigade, was reported to have "worked hard and so far as he was able kept the whole force under him working hard also." By late afternoon Darwent had personally directed the placement and operation of both of the city's manual water pumps. With the able assistance of Mr. Clark from the public waterworks supervising the connections to the water hydrants, the fire brigade (which at one point during the fire had 8,000 feet of canvas hose in use) had several thin streams of water directed at various portions of the fire—three such streams on Queen Street, three on Frederick Street, four on King Street, and three more on Chacon Street. The bluejackets from the *Buzzard* also brought a small pumping engine ashore. Yet nearly all observers agreed that the fire finally lost its strength, less from the fire brigade's and bluejackets' valiant water spraying efforts than from the luck and good fortune owing to the time of day. The wind calmed at sunset, allowing drafts of air to blow "inwards towards the flames and the fire began to burn itself out."[1]

Despite the fire brigade's hard work and the strategically directed streams of water, the city's water system had failed when it was needed most. Fluid had trickled, not gushed, from the hydrants because there was insufficient volume and pressure to provide any but the most anemic defense against the flames. An apologetic Governor Sir Frederick Napier Broome reported to London after the fire "that the water supply was most inefficient both in quantity and pressure." He also sent a detailed report of a public meeting on March 6 called by the city's mayor in which a relief committee was formed, newly unemployed were enumerated (219), and various plans were discussed both for aiding those directly affected by the fire and for adopting measures so that water might be available the next time fire broke out. Things were not uniformly bad, according to Broome. All of the local merchants seemed to have adequate insurance coverage for their losses; besides, "the Government will benefit by the duties which will be paid on the goods imported to replace those burnt," a fire-induced blessing that perhaps only a colonial functionary could truly appreciate.[2]

Speakers at the public meeting of March 6 bemoaned the poor, out-of-date equipment currently used by the city's fire brigade but focused their real attention and passion, and not for the first time, on the need for a more reliable city water supply. Action, not promises, was necessary from the colonial government. The problems of Port of Spain's rapid growth and its attendant needs for water, for drinking and bathing but certainly, in this case, for fire prevention, were issues that many of these same people had discussed before. They alluded to the terrible fire at the Hotel Guiria on Marine Square only a few years earlier; after that blaze many of the same people now in the room had discussed these same problems. Indeed, the hotel fire on February 15, 1891, had taken twelve lives, including several members of a visiting Venezuelan family. The report following the 1891 fire exposed water deficiencies very similar to those that had aggravated fire-fighting efforts again in March 1895. The Marine Square fire, just like the recent one, had been ineffectively combated by the "feeble jets of leaky hoses," and after that fire one newspaper had warned as to what might happen again. In part because of a poor water supply, Port of Spain, the newspaper suggested, represented "a population of near 50,000 souls . . . [that] goes to sleep every night on the brink of a volcano."[3]

One week after the March 1895 holocaust, the *Port of Spain Gazette* responded to public alarm by highlighting the need for a more efficient city fire brigade. Describing the modern equipment, organization, notification systems, and financial backing that supported the Metropolitan Brigade defending London from fire, the newspaper pointed out that Port of Spain, although smaller and poorer, would do well to emulate London insofar as it could. At the very least, Port of Spain needed one or more steam fire engines, new chemical fire extinguishers, and a proper system of electrical fire alarms. To pay for these improvements, the colonial and borough treasuries

Fire at the Hotel Guiria, Port of Spain, Trinidad, 1891 (from Gerard A. Besson, *The Angostura Historical Digest of Trinidad and Tobago* [Trinidad: Paria, 2001])

might be tapped and the local representatives of fire insurance companies operating in the city—of which there were more than twenty—might each be taxed £20 per annum. Most important, the report emphasized that none of these innovations and improvements would be worthwhile without "the all important one of a fully adequate supply of water."[4]

Water—or more explicitly, the lack of it—was an overriding concern to nearly everyone residing in the British Caribbean in the mid-1890s. A savage drought in the low-lying islands compounded the misery of the sugar bounty depression, reducing crop yields and especially punishing the working peoples in the region. In 1894–95 in parts of Barbados, the aridity shriveled subsistence crops and contributed to parish-level death rates of 44 per 1,000 "without the occurrence of any special epidemic of disease."[5] Conditions were even worse in Antigua during the same two-year drought. The public standpipes in St. John's were cut off except for short periods each day; in the rural areas, poor people "were kept on the roads at nights carrying small quantities of water from long distances." Plantation ponds in Antigua reserved for watering livestock were fenced and guarded, and desperately thirsty people were "stoned away" from these ponds when they approached for water.[6] In March 1897 in Jamaica, a terrible drought in the western parishes saw water available only from small hospital wells, and two rural Jamaicans were shot with firearms for attempting to take water from plantation-owned tanks.[7]

The Port of Spain water problem was different from the rural drought conditions elsewhere in the region. The low-lying and more northerly islands had always been well known as drought prone. Probably far more important to poor urban dwellers in the region in the 1890s, the availability of fresh piped water, which they fetched short distances every day from city standpipes, clearly delineated their recently improved existences from those of their kin who still resided in rural parishes and who continued to haul water, often for long distances, from unreliable tanks, wells, and ponds. For white residents of the region, a ready supply of fresh water represented an important symbol and amenity of modern life. This was, after all, nearly the twentieth century. Port of Spain, the eastern Caribbean's premier city, enjoyed direct steamer connections to North America and Europe, electric streetlights, a city tramway system, macadamized streets in most of the town, telephones and the telegraph, and many of the cultural and social events that accompanied modern urbanity. For these reasons and others, a reliable system of fresh piped water was considered an absolute necessity. As elsewhere in the world at the end of the 1800s, piped water in Caribbean cities not only underpinned and supported the concentrated presence of urban populations, but it also possessed a dynamic character because it attracted people from country districts to town.

In terms of access to piped water, Port of Spain was not far behind leading cities elsewhere in the world. London had rudimentary pipes and drains for carrying water in the eighteenth century; but many of London's poorer areas were badly supplied into the late 1800s, and not all London houses were connected with running water before 1899.[8] The Croton water aqueduct, completed in 1842 to serve New York, was one of the great American public works of the nineteenth century, and by the 1850s the "Croton bathroom" and a sewerage system were fixtures in the city; but some New Yorkers continued to rely on spring and cistern water into the late 1800s.[9] In the United States in general, although all of the largest cities had installed water supply systems by 1850, "particularly in the drier portions of the country, the delay was much, much longer."[10] In the southern United States, even in the large cities, piped water was by no means common until the 1880s and even the 1890s.[11]

In 1854 colonial engineers in Trinidad first directed water in cast-iron pipes from the Maraval River down to Port of Spain to supply the city, the first such innovation in the Lesser Antilles.[12] Within the next three decades, all of the British colonies in the region—except the Grenadines and the British Virgin Islands—installed gravity-powered, piped water networks whose principal arteries ran from upland reservoirs to the capital cities or towns. The locations and routed directions of piped water to certain places and not to others in each island created immediate and visible resource inequities that exerted a different kind of pressure on local colonial authorities. Everyone wanted the water; those in the country clamored for it, too. Urban dwellers who finally had it always seemed to desire a progressively more reliable and voluminous and cleaner supply. By the end of the century, in each of the islands new reservoirs were tapped and created and planned, new water mains laid, and other improvements made. Water now served daily needs for the elite and working classes alike. The presence of water in volume provided the means, as never before, to combat fire. Piped water was first introduced to Bridgetown, Barbados, for example, on March 29, 1861; the city's first tap was opened at Nelson's statue at Trafalgar Square and designated "for fires only."[13]

"Clear and Cold . . . Piped from the Reservoirs"

The unevenness of the provision of these first piped water supplies in the region, almost exclusively destined for plantations or urban areas, left rural peoples continuing to rely on the water sources they had always known, a reliance lasting well into the twentieth century in some places. At the turn of the century the water situation at the village of Gros-Islet, on the northern end of St. Lucia, underlined the combined problems of climatic variability and the kind of government neglect that accompanied geographical isolation, and it also provided an exemplary reminder of how everyone in the islands had secured water not so long before. In May 1901 Gros-Islet, to which a government doctor came only once per week, had no piped water, few nearby rivers, and unreliable, seasonal rainfall. So nearly everyone had a well in the yard providing ephemeral supplies of "not very wholesome" water. "There is also a large cistern hard by the R.C. church, which collects the water from the roof of that building. This cistern is kept under lock and key, and on certain days of the week the police . . . dispense water to the people."[14]

Proper sanitation and disease prevention depended on good water supplies, so the dearth of fresh water in rural areas created problems more important than simply the inconvenience or length of haul.[15] Although relationships between water supplies and disease incidence were imperfectly understood, in the rural parishes of Barbados in the 1890s—typical of other drought-influenced areas in the region—observers blamed a "want of pure water" for the prevalence of diarrhea, dysentery, and typhoid "hurrying many to an untimely grave."[16] Indeed, the terrible 1854 cholera epidemic in the British Caribbean, and its association with inadequate water supplies, likely hastened the earliest introduction of piped water to the region. The outbreak was particularly severe in Barbados, where an estimated 14,000 to 15,000 people died, many among the urban poor. Well water in Bridgetown in those dark days came mainly from Beckles's spring and was widely considered unfit for drinking. The quantity of Bridgetown water was, of course, lacking as well at midcentury. "Bathing was a luxury, and there was not enough water to clean the streets or the yards, to carry sewerage, or . . . to cope with fire."[17]

But the arrival of piped water in urban areas in the late 1800s did not necessarily eliminate contamination and the associated threat and fear of disease for city dwellers. An outbreak of dysentery in Dominica late in 1889 cast suspicion on the reservoir near Roseau "which furnishes the hydrants with water for the town."[18] In September 1892 an outbreak of cholera in Asia frightened the officials in Port of Spain and caused concern that the city's cesspits could contribute to a similar fate for Trinidadians. The following February and March saw renewed agonizing over the continuing public health problem in Trinidad's capital as the lack of available water prevented the cleansing of the city's public gutters and an associated reawakening of "the cholera scare of last year."[19]

Every public water system in the islands differed in detail from the others, although they all had fundamental features in common. An elevated reservoir or catchment area was fed by streams, springs, or man-made channels and connected by pipes leading down from this reservoir to the town or

Watering the streets in Bridgetown, Barbados, about 1900 (from Ann Watson Yates, *Bygone Barbados* [Barbados: Blackbird Studios, 1998])

final destination. At these destinations the recipients opened spigots or hydrants to fill receptacles. Improvements refined this basic model after the first such pipes were laid in the larger islands starting in the 1850s. In subsequent decades sand filter beds installed at intervening water stations in several places reduced the impurities in public water supplies. By the last years of the nineteenth century, the water systems had become sufficiently refined in most of the islands that progressively smaller pipes leading from water mains provided fresh water to individual buildings and even to private dwellings; then the introduction of water meters with associated schemes for paying for the fresh water led to an explosive sociopolitical issue in Trinidad at the turn of the century, with reverberations felt throughout the whole region. Because each island obviously depended on its own natural environment for its water supplies, periodic droughts, mainly in the northern islands, sometimes affected the reservoir levels and the public water supplies there. In 1903 an American residing in St. Kitts described Basseterre's supply as adequate for most of the year, "clear and cold" water being "piped from the reservoirs part way down." He further observed that the town's water was "locked off" part of the day during the dry season, a shortcoming that probably could be addressed by "properly constructed reservoirs."[20]

Overall control and responsibility for the establishment of local water systems varied slightly from one island to the next. In the smaller colonies the governments assumed initial control. In the larger places private water companies accomplished the first groundbreaking and laying of pipes, a jurisdictional decision modeled, perhaps, on London, which had several local water companies well into the nineteenth century. In Barbados the island's two private water companies became engaged in a dispute over the 1894 contract for supplying some of the island's rural areas. The Water Supply Company already had been providing water in some of Barbados's rural parishes for several years, but the Bridgetown Company bid for the rural contract as well, creating a dispute that soon spilled over into local personal and political rivalries. The disagreement eventually led the Barbados government to assume control of the island's entire water system in 1895, and by the turn of the century all of the British islands in the eastern Caribbean had centralized, government-coordinated water supply systems.[21]

Especially in the smaller islands, public water systems fed by necessarily finite highland reservoir areas were usually intertwined with the water needs of local sugarcane estates. When plantations and public agencies alike tapped common reservoir areas, both cooperation and conflict could occur, depending on different geographical characteristics and, perhaps more important, varying personalities. In 1883 the town of Sandy Point, St. Kitts, began to receive water that was shared with nearby Farm Estate. The water was transported in three-inch pipes from two mountain springs to a holding tank at the plantation. Then part of it was diverted to a reservoir just above Sandy Point. "From there the water is run into another reservoir situated at the Market Place and is distributed to the people once or twice a day."[22] This type of cooperation was nowhere to be seen in St. Vincent three years later. There, channeled water above Kingstown was diverted to the water-powered sugar grinding mills owned by the island's leading planter, Alexander Porter, whose antigovernment reputation and behavior had been the subject of fretful official colonial correspondence for years. During much of 1886, because Porter intercepted the water for his highland estates, it was available for only brief periods each day at the twenty-three standpipes serving the Kingstown populace. Eventually the governor of the Windwards had to travel from his home station in Grenada to purchase, on behalf of the government, water rights from Porter.[23]

The initial laying and burying of iron water pipes and the similar work subsequently necessary to extend, repair, and improve local water systems provided immediate, nonseasonal sources of wages for local men. Working peoples, moreover, invariably preferred these jobs to the usually more arduous plantation labor. In April 1890 the Barbados water companies were laying pipe "in every direction and paying good rates" in outlying St. John parish; "but laying pipes will not be a protracted service, and then cane cutters we hope will be numerous."[24] Rural public works projects were performed out of sight of many island residents and therefore inspired little criticism, but such projects became more immediate sources of comment when they occurred under the watchful eyes of urban taxpayers. The installing of new water pipes along the streets of Castries, St. Lucia, early in 1897, a slow-moving operation hindered by poor administration and supervision, involved interminable idling and a "scandalous waste of time." It was enough to cause speculation about time wasted "during the many months that hundreds of men have been employed in the bush out of sight of criticism."[25]

Once town and city standpipes were installed, these points automatically became gathering places for urban dwellers, who brought pails to fill, a competitive daily process that involved more than a little jostling, pushing, and commotion, especially when water was available only for limited periods. Like West Indian marketplaces, public standpipes in Caribbean towns and cities not only served economic functions, but they also represented sociocultural focal points. What was taught and learned there, however, did not necessarily please everyone. Early in 1885 only two standpipes served many of the residents in one neighborhood of Castries. Local women were described as crowding and even fighting over the water available there, using "the most disgusting language . . . closely imitated by the still larger crowds of children of all ages. . . . It is sickening to hear the abominable expressions screamed out by little girls who can hardly speak."[26] Similar reports told of occasionally desperate crowds scuffling around rural standpipes when water supplies eventually came to country areas of the eastern Caribbean. Yet the first water piped to rural zones nearly always was considered an unalloyed boon by the villagers there. All rural peoples in Barbados in 1897, for example, seemed pleased by the extension of "perfectly pure" water into the countryside, possibly because unlike water provided for individual houses, the public water from "Queen Victoria's Pump" came free of charge.[27] In some isolated and drought-prone parts of the region, piped water could mean, very simply, the difference between life and death. In emphasizing the need for a reliable water supply for the cane workers' villages on the northern end of St. Kitts, the district medical officer remarked that "during the drought . . . of 1882, the deaths in the Parish of St. Ann were many more than during the same period in 1883, when from the abundant rains, the inhabitants were enabled to supply themselves with wholesome water."[28]

The coming of piped water to the Lesser Antilles represented a common regional experience in the British Caribbean similar to changes occurring elsewhere in the empire in the late 1800s. Rough comparisons may be drawn between colonial officials with expertise in water engineering and the cadre of tropical forestry experts serving the empire. Both groups depended on and shared specialized published knowledge housed in and disseminated through London, and they augmented this common technical knowledge with their own on-the-spot experiences. Not unlike some tropical foresters, water engineers traveled throughout the empire, dispensing advice and expertise in exchange for monetary stipends paid from local treasuries. Captain Alexander, for example, provided helpful directions in maintaining proper water systems for the governments of Grenada, Antigua, and St. Kitts early in the 1880s, so the water company in Barbados requested that the local government invite him to address insufficiencies in Bridgetown's water supply.[29] Yet water engineering was sometimes considered a more general skill than forestry, so it was often handled, or at least coordinated, by individuals with backgrounds less specialized than those of scientific botanists. For example, when a new reservoir and system of water filter beds were being considered for Castries, the advice and expertise of the St. Lucia administrator, Charles Anthony King-Harman, was considered helpful because of his somewhat recent experience with installing the waterworks in both Cyprus and Mauritius.[30] Somewhat similarly, George Blanc, a white Dominican, was educated in Europe as an engineer and returned to his home island to build two bridges and to supervise the installation of the system of water pipes there. He also supervised the construction of new water systems for Montserrat and parts of St. Kitts. Before his death in 1893, Blanc also had served, at various times, as a member of the Dominica Legislative Assembly and the Board of Health and as a district magistrate, playing parts in a multiplicity of administrative roles that were very common for white men of the smaller islands.[31]

The bringing of piped water to the eastern Caribbean also introduced a previously unknown series of technical problems that could, on occasion, provide more harm than help. Examples from Dominica, just after waterworks were installed at Roseau, illustrate the point. In 1880 water brought down to the town from the mountain reservoir ran from the open pipes and then simply flowed through the streets. Water had thereby been "supplied," but it was as if Roseau itself was being poured into a river, as its sodden streets and lanes were in a continuously semiflooded state. The subsequent introduction and installation of "patent spring corks" improved the situation somewhat. Water now flowed from the pipes when the corks were pushed, "which completely prevents the constant flow of water from which we formerly suffered." Yet the spring corks were neither numerous nor completely effective, and inhabitants of Roseau had to contend with the artificial provision of far too much water, at least into the following year.[32] Discussions about providing fresh water for the nearby settlement of Point Michel were under way in 1881, with elaborate plans for extending two-inch cast-iron water pipes from the hills above. But the project still was not realized four years later; the long-awaited pipes had indeed arrived by steamer from London, but they were only one-inch pipes, "the result of a clerical error."[33]

Technical and administrative malfunctions also occurred in the other islands. In Antigua, where it seemed there was always too little water, a water pipe burst, flooding the streets of St. John's in August 1890, "no doubt owing to the heavy pressure."[34] Piped water first came to Grenada in mid-1879, and by 1886 in St. George's the water consumed there for drinking, washing, cooking, and fire control underwent a filtration process at the local reservoir—or so residents of St. George's thought until June of that year, when "an eel, and especially a large one, [was] seen to issue from one of the fire-plugs in town." The surprise of seeing a full-sized eel slither from a pipe carrying supposedly pure water led one perceptive journalist to report that "all the water that comes to us in the town, although supposed to be filtered, does *not* pass through the filtration process at the Reservoir before entering the pipes."[35]

When these and similar problems occurred in Caribbean towns and cities, it was only natural for residents to seek remedies from officials or, perhaps more often, to blame those in charge of the new water systems. The furor over the city water supply's shortfall after the Port of Spain fire in 1895 was the kind of public discussion and faultfinding that played out in the smaller islands of the region in many different ways in the latter years of the 1800s. The Grenadian eel, for example, had to have set tongues wagging about the indolence and incompetence of Charles Risk, the official in charge of St. George's public works. Only the year before, the entire town had been without drinking and washing water and subject to fire damage without recourse for three consecutive days, a period during which Risk had been incapacitated from drunkenness.[36] An apparently disagreeable official in charge of St. Lucia's waterworks in the following decade came in for roughly similar scorn. H. S. Osment was on obviously poor terms with leaders of the Castries fire brigade in 1898, reluctantly turning over equipment to them for a fire drill in late June. When the fire brigade members "blew out a joint" on the town's water main, Osment ranted to the governor that the incident was the fault of "unqualified valvesmen . . . and caused a large section of the town to be thrown out of fire-protection during the Corpus Christi Festival . . . when the population of Castries was at its maximum." Yet Osment's technical qualifications were held in generally low regard, and when he built a small structure adjoining the town's reservoir for £300 to house three water meters in 1899, he was condemned as a typically arrogant colonial functionary and squanderer of public funds.[37]

Looking to obviously fallible colonial officials, rather than to cloud formations in the heavens, for solutions to local water problems marked an important shift in the way nearly everyone residing in the eastern Caribbean in the latter half of the nineteenth century contemplated and coped with subsistence and survival. Water was, after all, "life's true and unique medium" without which "life simply cannot be sustained."[38] But it was no longer the precious fluid coming from local springs, cisterns, and rainfall and thereby routinely obtained at the family, village, or city well. Rather, water was now owned, preserved, fenced, channeled, piped, and rationed by an impersonal system operated by human beings. To be sure, piped water systems that brought filtered water supplies were far more convenient and hygienic. But the price of convenience and purer water was a dependency on public standpipes installed and maintained by a hierarchy of often faceless workers and supervisors. Moreover, a growing and concentrated urban population did not simply need a piped water system; their very presence depended on it.

When water systems failed, as they did for any number of reasons, including drought, mechanical breakdown, poor maintenance, inexperience, or mismanagement, the urban poor and, increasingly, those in rural areas now had few other places to which to turn. They knew the local men who had laid the pipes, but they were obviously not acquainted with, and perhaps had only seen at a distance, the colonial officials in charge. Water problems therefore came to be classified and contemplated as problems created by the "government." It is worth suggesting that the advent of modern water supplies and other amenities in the region at the end of the nineteenth century at once improved local living standards yet also reinforced government as an amorphous yet localized and common target when things went wrong. In the earlier days, plantation owners and overseers decided one's wages and much else. Of course, these men continued to be powerful well into the twentieth century, often working and, in many cases, being interchangeable with government officials. But now, especially in towns and cities in the eastern Caribbean, the resources and opportunities that were the stuff of peoples' daily lives were becoming entangled in an increasingly complex web where interrelated social, economic, and political decisions were made by individuals they never saw.

The impersonality of modern water systems in the islands did not mean that locals were ignorant of the general schemes, layouts, and networks of reservoirs and pipes. Indeed, commentary on the conservation and protection of forested upland watersheds took on an obviously practical, not simply a protectionist or aesthetic, tone. The editorial in the *Port of Spain Gazette* early in 1889 likening a forested watershed to a "great vegetable sponge" deplored the deforestation in the vicinity of the Maraval Valley reservoir above the city and the apparently widespread violation of the legally specified thirty-foot forest buffers along either side of watercourses in Trinidad's hills and mountains.[39] Protection of the forests surrounding the streams above Port of Spain was considered vital because the city's water supply hung in the balance. It was not the first time such concern had been publicized. In 1885 the same newspaper had bemoaned the cutting of branches and grasses for stock forage along the streams above the city. "How all this happened in face of an Express Ordinance . . . preventing cutting any wood near the borders of a river is strange indeed. Where are the Ward officers?"[40]

This point had to be driven home not only in Trinidad but in the nearby islands as well. Accordingly, the need for protecting upland watersheds in light of recently centralized water supplies throughout the region became a general theme appearing often in British Caribbean agricultural extension reports whose authors, in turn, were familiar with similar conditions elsewhere from the imperial forestry publications of the time. The interrelationships between forest conservation and water supplies also were stressed in the reports from the superintendents of the islands' local botanic gardens. Even in Dominica, the highest and the rainiest of the British Caribbean colonies, watershed conservation needed to be practiced and therefore monitored. The Dominica newspaper, in any case, saw fit to republish the text of an address by Daniel Morris to the London Chamber of Commerce in March 1888. In commenting on the recent regional forestry reports submitted by E. D. M. Hooper, Morris emphasized that West Indian forests were valuable "not only as reserves of timber . . . but as also means of maintaining the humidity of climate, and protecting the sources of springs and rivers." Morris concluded by advocating the taking of "measures which are best adapted to prevent extensive and reckless cutting down of forests necessary to their well-being and future prosperity."[41]

Important as the coming of piped water to the Lesser Antilles was for economic and health reasons, few published articles or reports about its introduction were silent about the vital issue of local communities finally being able to fight fire with an adequate volume of water. Even in the arid northern islands where recurring drought desiccated crops and drove people to brackish water holes, spokesmen acknowledged the role piped water would now play in "the all important question of protection against fire."[42] In earlier years towns and cities had husbanded small amounts of cistern water for combating fires, a strategy with dangerous limitations. When city areas suffered fire damage before the availability of piped water, as Bridgetown did in the late 1850s, the long-term results often were simply unoccupied "burnt areas" of rubble interspersed with hardy plants, zones not built back up for fear of destruction by a future fire and visible evidence of an inability to cope with such an event.[43] Before piped water, buildings at the ocean's edge could sometimes be protected with seawater, but that strategy extended only as far as the longest bucket brigade. Nonetheless, places where piped water came late still had to rely directly on the sea for fire control into the twentieth century. In July 1904 a general fire at Hillsborough, Carriacou, scorched the almond trees at the town square and caused only minor building damage because the town's residents formed bucket brigades, carrying seawater to fight the flames late into the night.[44]

As the availability of piped water helped to nurture and encourage ever larger populations in the cities and towns of the Lesser Antilles late in the 1800s, these urban places needed progressively more reliable water supplies to fight the inevitable fires that fed on the flammable building materials now crowded together there. This circular relationship, so obvious from the accelerating concern and commentary in local newspapers of the time about the need for progressively better water supplies, heightened local dependency on the adequate functioning of water systems so the effects of the kind of catastrophe that visited Port of Spain in March 1895 might be avoided. Thereafter, adequate volume and pressure from city and town fire hydrants provided satisfying evidence of how modern methods could combat the ravages of accidental and potentially destructive fires. When Mrs. Fonseca's house on Redcliffe Street in St. John's, Antigua, caught fire in October 1893, the town's fire brigade appeared quickly and extinguished the blaze with strong jets of water, providing "another proof of the efficiency of the service and the value of the water works."[45] But when the water system failed, as it often did, recrimination and blame ruled the day. When a hotel burned down in St. George's, Grenada, late in 1882, the flames had their way for so long that nothing was left of the structure, even though firefighters had arrived quickly on the scene; "but, as usual, an abundant or even sufficient supply of water could not be obtained on time—for half an hour at least."[46]

Once the city water systems were in place, even though piped water was considered novel and new for years, there was no substitute for dealing with the possibility of fire in the region's growing towns and cities. In January 1880 heavy rains in St. John parish, far to the east of Bridgetown, Barbados, created landslips and broken water pipes and a cessation of the water supply to the city. Directors of the water company arranged to have carters bring puncheons of water to town for drinking, and those with cisterns generously shared with friends and neighbors. But the real problem was a dread of fire before repairs could be accomplished during this normally dry period of the year. To make things worse, one newspaper complained that "the police have done everything in their power to make the people aware that the reservoir is empty."[47]

Why fear the rapid dissemination of news about a failed water supply to an urban populace now dependent on obtaining the fluid from local standpipes? The answer lies in part in the dread that colonial powerholders felt toward the local people they governed and controlled and the allied threat of arson with the intent to create widespread damage by fire. Only two years after the disruption of the Bridgetown water supply because of broken pipes, a serious fire of mysterious origin destroyed much of a large downtown store. Owing to a generous supply of water, a very serious city fire that "might have been a severe calamity to . . . Bridgetown" was prevented.[48] But given the random disruptions in supply and periods of low water pressure in city water systems, there was an almost continuous fear of water outages followed by fire, especially the premeditated, man-made variety. In June 1884 Bridgetown residents again went without water because of a burst water main on Roebuck Street. And again the dread of possible arson occupied the thoughts of officials, merchants, and planters: "The mischief which might ensue to property in our densely packed town is almost incalculable, and the fact of our large City population being cut off from their water supply is one which cannot be contemplated without the gravest apprehension."[49]

Fire Engines and Fire Brigades

The arrival on Saturday, November 11, 1893, of the British steamer *Atlantis* at the Castries harbor (recently enlarged to become the major coaling port in the southern part of the British Caribbean) was greeted with more fanfare and enthusiasm than usual because of the important equipment aboard. The freighter was bringing to St. Lucia a new steam fire engine, with its fire bell, ladders, and ropes, that had been ordered by the Castries Town Board from the Fire Appliances Manufacturing Company of Northampton, England. The smart-looking device consisted of a steam pump mounted on top of a boiler, and these gunmetal and brass components were attached "for conveyance purposes" to a high-wheeled wooden carriage. The bright red carriage body also carried a toolbox mounted on top that already bore the prominent, black-stenciled words "Castries Volunteer Fire Brigade—1893."[50]

The testing of the new fire engine, which took on the form of a public gathering and demonstration, came the following Saturday afternoon. At five o'clock the machine was rolled to the fountain at Columbus Square in the center of town. After the boiler was lit, the steam engine activated the pumping device, and the suction hose was placed into the fountain's pond. Soon two jets of water more than 100 feet high shot into the air from the canvas hoses, even higher than the surrounding palm trees, whose branches and leaves "yielded and drooped beneath the weight of water." The assembled town board members and other local dignitaries were joined by a large crowd of onlookers congregated "on the broad alleys under the trees," and the demonstration also was witnessed from nearby windows and verandas. Occasional cheers accompanied the jets of water as they arced into the sky. After the successful and exciting demonstration of the town's new fire engine, which lasted for a full hour, the *Voice of St. Lucia* wrote that all that was now necessary was manpower; it was hoped that the organization and training of an efficient fire brigade would encourage "our smartest young men" to volunteer to protect buildings, property, "and perhaps the lives of those who are dear to them."[51]

As the numbers of both buildings and inhabitants in the eastern Caribbean's population centers grew, and with the increasingly widespread use of both kerosene and wooden matches, all of the islands' governments recognized the need to improve and to build upon the informal but now old-fashioned means of fighting fires that had been in place since the earliest towns and plantation settlements in the region. Over time the traditional, informal methods of fire control had gradually been augmented and eventually replaced by firefighting institutions. During the 1780s the "friendly fire company" of St. John's, Antigua, had been formed as a result of the devastating fire there in 1782, and all members thereafter were expected to maintain in good order "a certain number of buckets and fire-bags."[52] At roughly the same period in Barbados, a number of Bridgetown fire companies were established, some of which used as firemen male slaves "wearing blue jackets and red capes with the initials of the owner . . . painted on the back."[53] Of course, the impromptu and hastily organized firefighting gangs of plantation slaves in rural areas had been the forerunners of the *cannes brulées* celebrations.

By the late nineteenth century all of the urban areas in the British Caribbean had paid fire brigades, or at least formalized institutional means of combating fires, that were the results of local legislation, court actions, and varying degrees of volunteerism. In the islands with smaller populations, the fighting of fires was usually considered one of the duties of the police, an arrangement that pleased very few. In 1883 in Dominica, for example, nearly everyone agreed that Roseau, rather than relying on the police to save property and lives in case of a major fire, was in dire need of a "proper" fire brigade.[54] Six years later in St. George's, Grenada, the maladroit handling of hydrants, couplings, and hoses by members of the local police force during a fire at night provided cause for gratitude among the populace that the wind had been calm that evening; further, the incident "proved to demonstrate the necessity of a Fire Brigade in St. George's."[55] In Castries in April 1898 the town's police constables, who were mainly Barbadians, were, as usual, jeered and obstructed by the locals when they turned out to combat a minor fire.[56]

The (usually negative) opinions published in British Caribbean newspaper articles, editorials, and letters about the competence, training, and composition of local firefighting personnel intensified toward the end of the century, especially in the weeks and months after well-publicized fires devastated local cities or those in neighboring islands. After the terrible fire of June 1890 destroyed much of Fort-de-France, Martinique, reports circulated about the failure of that city's water supply during the fire, as well as the relatively poor showing of the local fire brigade. Afterward, the expression of a feeling of urgency in, among other places, Dominica about the need to form a proper fire brigade—not just police taking on extra duties—emphasized volunteerism. Perhaps more important, Dominica was said to need a certain "class" of man to come forward to help combat fires because

A late nineteenth-century fire station in Bridgetown, Barbados (from Ann Watson Yates, *Bygone Barbados* [Barbados: Blackbird Studios, 1998])

the "functions of a Fire Brigade and those of the police are entirely distinct" and the "Police Force [should not] be the nucleus of the establishment."[57] Finally, in mid-1898 a "proper" volunteer fire brigade, composed of "respectable" members of the Roseau citizenry, held its monthly meetings and parades under the supervision of Police Inspector James.[58]

An assessment of fire brigade volunteerism and an awareness of the composition of fire brigades themselves in the eastern Caribbean in the late 1800s help to throw light on some of the region's social complexities of the era. Specifically, fire brigade service provided opportunities, in Dominica and elsewhere, for men with "respectable" backgrounds to distance themselves from the impoverished masses, especially the objectionable individuals who made up the rank and file of local police constabularies. Respectable identity might be attained through economic success or brown skin, and in a broad sense the term might be considered a euphemism or shorthand for light skin color. As historian William Green has suggested for the British Caribbean of the late nineteenth century, "Blacks were not excluded . . . but in island societies which were predominantly black, comparatively few people of pure African origins could be numbered among the so-called 'respectable' classes."[59] Police constables, in contrast, were hardly respectable; rather, they were often considered ruffians without scruples who had sold out to the colonial elite and were only too willing to push people around. In Trinidad a "deeply rooted antipathy to joining the police force" perhaps went back to the days of slavery when the free colored, or so-called Alguazils, were forced to become police as a mark of social inferiority.[60] Also in Trinidad, as well as in Grenada and St. Lucia, the itinerant Barbadians who composed the majority of police forces were often despised by local populaces for their cruelty and lack of civility. Throughout the islands in the late 1800s police sometimes were veterans of the West India Regiment, and it was not uncommon for these men, who had barely escaped criminal prosecution in their home islands, to become police constables elsewhere. These points perhaps help to explain why, in early 1881, a crowd in Port of Spain took particular delight in cutting the water hose that was being used by the police in attempting to extinguish a fire at the city's police headquarters.[61]

With the region's largest and most culturally diverse human population, Trinidad perhaps exemplified best how the composition and activities of a late nineteenth-century city fire brigade mirrored some of the Caribbean's social differentiations at the time. The colony's Special Ordinance number 17 of 1860 originally called for a volunteer fire brigade for Port of Spain. By 1900 the city brigade, under the command of Captain Darwent, was composed of two lieutenants, thirty-nine volunteers (many with Portuguese surnames), and "twenty paid men."[62] The latter, although they were full-time firemen, normally were relegated to manning the city watchtower, maintaining and cleaning station equipment, and dealing with day-to-day activities much as police constables did. Although overlap between the different classes of firemen was not uncommon, the real fire fighting might be best left to the volunteer fire brigade. A slow-moving and ineffective response by the city's paid firemen to a house fire on Queen Street in August 1892, for example, was quickly rectified by the reported alacrity shown by the volunteers when they reinforced the regular brigade and thereby saved several nearby houses.[63] Not everyone always agreed that the volunteers outshone the paid men, however. In 1892 Captain Fortescue, then the head of the Port of Spain fire brigade, preferred the paid men to volunteers, asserting that the latter were flabby, poorly disciplined, and maintained "the absurd idea that the paid men are their servants and treat them with great insolence."[64]

The planned social and ceremonial activities of the Port of Spain Volunteer Fire Brigade underlined the organization's exclusive character. Competitions in uniform inspection, rope climbing, running drills, and hose and ladder assembly were scheduled and advertised, with some of the city's leading doctors, merchants, and other dignitaries as invited onlookers and even acting as judges and timers. Dancing and iced drinks often followed the festivities. The governor himself sometimes sponsored the events, such as in October 1887, when a prize competition between the Port of Spain and San Fernando volunteer brigades at Brunswick Square was sponsored by Governor Sir William Robinson, who attended the festivities with his wife and daughters.[65] Officials of nearby islands took note of the gaiety, publicity, supposed effectiveness, and exclusivity of the Port of Spain fire brigade while they assessed their own immediate fire control strategies. In Castries in December 1894, the town board's fire committee considered whether they might be better off with a volunteer brigade composed of "owners of property and men having a stake in the Town." Their model, as far as social proprieties were concerned, was located nearby: "The Volunteer Fire Brigade of Port-of-Spain, Trinidad, is so exclusive as to admit candidates for membership only by election, as to a club."[66]

Of course, part of the city fire brigade's special social character was cross-cultural in nature and not necessarily tied to a Caribbean setting. The mystique, bravado, attention to uniform appearance, and even selectivity of members all have been elements of historic fire brigades elsewhere, such as in Britain at the time as well as in the United States in the late 1800s. Social standing or at least good character has also counted for fire brigade membership in places other than the Caribbean where members are drawn "from among leading citizens and respectable artisans" rather than "rowdies."[67] Even some of the self-congratulatory identity and terminology for Trinidadian firemen of the late 1800s was the same as that used elsewhere. Two weeks before Christmas 1894, a small fire on Frederick Street in Port of Spain, as a local newspaper of the time pointed out, posed little problem for the volunteer "fire laddies" who were reported to have arrived punctually and worked efficiently and bravely without regard for their own personal safety.[68]

Some of the safety and evacuation equipment in use in Trinidad at the time would probably be recognizable to fire brigade members elsewhere one century later. On the evening of September 6, 1900, at the government Red House in Port of Spain, Captain Darwent led select members of the city fire brigade in demonstrating some of the group's latest rescue equipment. Ladders were extended all the way to the top of the building, from where several brigade members slid safely to the ground through a long, funnel-shaped, canvas apparatus. Then several brigade members demonstrated the use and deployment of "jumping sheets" by leaping from a height of twenty-five feet onto the taut canvas held by others. After these demonstrations, which drew a large public audience as usual, the brigade members entertained with music and singing, and a dance followed in the recreation hall of the fire station.[69]

Despite the continuing publicity for the social niceties and respectable character of fire brigade membership in Port of Spain, it is likely that the city's huge 1895 fire reduced enthusiasm for ceremony in favor of fire protection at any cost. Accordingly, the city's first steam fire engine, which was to replace the old and worn-out hand pumps, arrived in March 1896. The machine was manufactured by Shand, Mason and Company of London, could pump up to 30,000 gallons per hour, and was pronounced "in every way well-adapted for its uses in Trinidad." This exuberant boosterism concerning the new city fire engine, reminiscent of the similar excitement three years earlier in St. Lucia, continued two weeks later with a similar public demonstration of the engine's ability to pump water into the air and thereby entertain crowds of onlookers. Indeed, the new Trinidad fire engine continued for some time to provide public spectacles. On Saturdays at prearranged locales and times around the city it could be seen spraying its jets of water skyward to "manufacture rainbows" for the delight and amusement of groups of adults and children.[70]

The new steam fire engines, in Trinidad, St. Lucia, and the other islands, required a new level of technical sophistication (not to mention ready supplies of coal) for their operation and maintenance. The machinery and its boilers also supplanted in part the manual carrying of water and the use of hand pumps to fight fires as in the past. The first steam fire engine probably was invented in London in 1829, and by 1866 fifteen American cities had adopted steam in order to "fight fire with fire."[71] But the steam firefighting engine's relatively late arrival in the Lesser Antilles augmented rather than replaced traditional firefighting techniques and equipment, and Caribbean steam fire engines were confined almost exclusively to urban areas and to places on good roads. "The Report on the Fire Brigade for 1899" for Barbados enumerated a steam engine, hand carriages with hose reel attachments, nozzles, fire axes, scaling ladders, ropes, and other items among its apparatus. "Horse hose reels" were also important equipment, and detailed expenses for the purchase and upkeep of horses were major items in the published budget.[72]

The new firefighting innovations at the turn of the century included chemical fire extinguishers imported from the United States. The Racine Fire Extinguisher was purchased by private companies as well as public fire brigades in the Lesser Antilles by the turn of the century. The cylindrical, and by current standards quite recognizable, Racine put to practical use "the carbonic gas idea," and it came in different sizes. The hose attached to a container with a capacity of 3 to 5 gallons could project a stream of water—apparently a light mixture of carbonic acid—from thirty-five to fifty feet. A "wheeled machine," on the other hand, with a tank of up to 100 gallons, had a "practically unlimited" range for its jets of fluids. Part of the attraction of the Racine was its advertised life span of up to twenty years without the need for recharging. The claims for the fire quenching abilities of this machine notwithstanding, a number of businessmen, plantation owners, and civic leaders had, by 1900, purchased the device so as to reduce the premiums for their fire insurance policies.[73]

Issues concerning the social composition, equipment, and overall competence and training of city fire brigades in the small islands of the eastern Caribbean at the turn of the century posed legitimate and important problems in preventing possible urban fire disasters. Residents of smaller towns and rural areas, however, usually had little or no organized fire protection. Plantation owners hastily organized their own laborers to fight fires on the estates. But if fires occurred elsewhere, and if owners or residents themselves were unable to extinguish the blazes even with the help of a few neighbors, rural buildings often simply burned to the ground. Most of the structures at Codrington College in Barbados burned down in April 1885 at the end of a long dry period. Nearby laborers tried to help with little effect, and a detachment from the Bridgetown fire brigade arrived by train hours after the fire had done its damage.[74] Three years later a house fire in Sandy Point, St. Kitts, burned down without so much as a visit from the local police, leaving the residents wondering why taxes were assessed on every "miserable little hut" in their town, as well as in Dieppe Bay, if they received no fire protection.[75]

Fire Insurance

The availability of fire extinguishing machines such as the Racine, the perceived efficiency of local fire brigades, and more than anything else, the quality of piped water systems in the British Caribbean all played interrelated roles in the calculation of local fire insurance rates. These fire insurance premiums, in turn, represented important costs of doing business and even influenced the availability of merchandise in the region at the end of the nineteenth century. Since the local agents representing British fire insurance companies usually held other commercial positions and, often in the smaller places, official governmental posts, questions of fire insurance availability, coverage, and rate structures often came to the fore in discussions and meetings at the highest local levels. As one of many examples, an exchange in the Barbados House of Assembly in mid-March 1883 about recent fire-related correspondence between local fire insurance company agents and the island's governor naturally led to a discussion about improvements in Bridgetown's water supply system and the possible additions of new hydrants and pipes.[76] On occasion, considerations of fire insurance even transcended individual islands to become genuinely regional issues, such as the 1898 newspaper warnings of prohibitive fire insurance rates delivered by the Kearton company.

British fire insurance, although a historical latecomer when compared with maritime insurance in England and elsewhere, traces its origins to the seventeenth and eighteenth centuries; its beginnings were evidently encouraged by the great London fire of 1666.[77] As the British insurance business evolved in general, fire insurance companies extended coverage to foreign areas as well as to overseas colonies in the British Empire, including the Caribbean. As early as 1782 the Phoenix fire company of London insured stores and warehouses in Antigua; before then, "fire insurance did not extend to this island."[78] But by the end of the nineteenth century, every British colony in the eastern Caribbean had at least five or six British insurance companies that regularly advertised their presence in the region, their longevity (e.g., Sun Fire Office London, Established 1710), and their availability to cover losses in buildings, property, and crops in return for the payment of annual premiums. Trinidad, and especially Port of Spain, had the greatest number of insurance companies writing local fire policies. After the city's 1895 fire, local merchants and property owners filed aggregate losses of more than £350,000 with twenty-two different British insurance companies.[79]

The absence until early in the twentieth century of locally based mutual fire insurance companies in the islands compounded the imperial domination of the British Caribbean economy. In a more tangible sense, this void was considered a factor in reducing local economic growth because insurance premiums accrued to Britain and were not reinvested in the islands themselves. An 1886 estimate for Trinidad reckoned that $30 million in local property was insured by British fire policy companies. (Sums were routinely compiled and reported at the time in both British pounds and Caribbean dollars.) Annual Trinidadian insurance rates, moreover, varied from 0.5 percent to 3 percent of the estimated value of the property insured. If the average fire premium on $30 million was 1 percent, the estimate continued, one could assume that, even after local agents' salaries were paid, close to $300,000 was "annually sent out of the colony in fire insurance premiums" and thereby unavailable for local reinvestment because of this "flight of capital."[80]

The notable presence and success of the Barbados Mutual Life Assurance Society, established in 1840 and thereafter playing an important role in local Barbadian capital formation and investment, did not have its nineteenth-century fire insurance counterpart anywhere in the islands.[81]

Two indigenous fire insurance companies were formed in British Guiana, one in 1865 and the other in 1880, and although serious discussions of local fire insurance company formation took place in Barbados early in the 1880s, no such venture was realized in the Antilles until early in the twentieth century in Port of Spain. The Trinidad Co-operative and Mutual Fire Insurance Company was formed in May 1904, with its stated intent of lowering local fire rates that had been increasing after the 1895 fire and because of subsequent altercations over the city water supply. The directors of the Trinidad insurance company intended to finance the operation with local promissory notes, vowing to keep rates within reason and thereby helping to reduce the capital flow to London. Later in the year the Port of Spain paper reported city fire premiums to be a "trifle lower" owing to a new "spirit of competition," and at the end of 1904 the local company directors were described as being "almost philanthropists" in their service to the community, suggesting that their finances were probably in shaky condition.[82]

Although fire insurance coverage and premiums were considered mainly urban issues, they played important roles in some of the region's rural zones as well. In turn-of-the-century Barbados, sugarcane fires of course precipitated discussions between government officials and insurance agents so as to raise eyebrows in London. In some of the other islands, owners of sugarcane and cacao estates insured only their dwellings or estate buildings, paying rates that were usually lower than in urban areas. During the sugarcane harvest season in St. Vincent in 1881, an unpopular estate manager was targeted by "a certain class of laborers" who vented their anger by lighting a series of "megass house" fires, leading to rewards being offered by the "Insurance Office."[83] But the real presence and influence of insurance companies were not in the country districts, but in the region's towns and cities, where fire insurance agents advertised rewards for the conviction of incendiaries, made shows of publicly thanking soldiers and fire brigades for firefighting heroics, and with police officers and other public officials, interrogated witnesses at postfire investigations.

Fire insurance policies in the region covered the structures and contents of businesses, government buildings, homes of the well-to-do and middle class, and many of the small shops in towns and cities. Whereas one might interpret the taking out of a fire insurance policy as an individual's attempt to reduce risk and uncertainty, spokesmen in Trinidad in the mid-1880s termed fire insurance arrangements "gambling policies," although they conceded that such a gamble made sense during periods of city fires.[84] From the opposite point of view, the company of course gambled that the insured would not burn his or her own property. A cynical passage from an 1896 Barbados newspaper about Trinidad underlined the point: "It occasions no surprise now when we are informed by Cable that another fire has destroyed several stores in Port-of-Spain, Trinidad. . . . Covered by Insurance of course, and origin of fire unknown of course. The Insurance Companies will have to protect themselves against these frequent illuminations at their expense."[85]

Uncertainty might be reduced for the fire policy holders if they conducted business with reliable companies, but as in the United States at the time, seekers of low insurance rates sometimes purchased from companies with inadequate resources, leading in turn to disastrous results when fire damage occurred.[86] The news of hundreds of unpaid insurance claims for houses and shops after the Fort-de-France, Martinique, fire in 1890 unsettled urban dwellers throughout the British Caribbean. Of the estimated 1,500 insured properties destroyed in the Fort-de-France holocaust, as many as 1,200 were reckoned a "dead loss" to their owners because they had been insured with the Mutual Insurance Company of Martinique, a company with only limited capital reserves and that could meet only about 10 percent of its obligations after the fire.[87] Similarly, after the 1907 earthquake destroyed downtown Kingston, a number of the insurance companies doing business in Jamaica "disclaimed responsibility" for the terrible fire losses, invoking the technicality that the insured did not have coverage for the earthquake that had originally caused the fires.[88]

A principal reason for the presence of less-than-reliable firms doing business in the Caribbean was that property owners subscribing to fire insurance throughout the islands never tired of complaining about high premiums, and they therefore sought cheaper fire policies. Especially in the smaller islands, owners of both commercial and private property lamented the high "English" or "European" fire insurance rates and professed envy of those in British Guiana who reportedly paid less because of the local company there. A report from Dominica in 1903 bemoaned fire insurance premiums that were "so high as to be almost prohibitive." As a result, very few of the local houses or even public buildings there were said to be insured. The report continued that Roseau householders were expected to pay from 2 to 2.75 percent of the value of their property annually for fire insurance protection. "It is therefore not to be wondered . . . that owners refrain from insuring their property."[89]

The importance of local fire insurance rates and the conversations surrounding them cannot be overemphasized in any attempt to understand what business conditions in Port of Spain were like at the turn of the century. By most accounts, the commercial fire rates doubled between 1895 and 1901; in some cases they more than doubled.[90] The increased rates, in turn, led to considerations of the relationships between local agents and British insurance agencies. The increases also called for a serious and apparently honest assessment of the city government's fire prevention infrastructure, especially the existing water supply, as well as the laws and regulations for the city and its future. The higher fire rates even inspired squabbles among merchants about which parts of the city were more dangerous than others. Further, the increased premiums led store owners and consumers to wonder about the quantity and quality of imported goods they could expect to provide and purchase in local stores and shops. Discussions of increased fire rates in an atmosphere clouded by the continuous threat of fire placed blame, sought culprits, and even engendered musings about the morality of the city's populace. In short, considerations and commentary about local fire insurance rates illuminated contemporary thinking about what the city represented and where it should be headed.

In some instances, commercial fire rates doubled in less than a year after the big 1895 fire; the Lancashire Insurance Company raised its Port of Spain rates to 30 shillings per £100 worth of merchandise by February 1896.[91] The company's agents, responding to local outcries, sought rate reductions from their British correspondents by pointing out that the city had appointed a new fire brigade instructor, purchased a new fire engine, enacted stricter building laws, and widened Frederick Street by twelve feet. In the next year the high rates led to widespread worry that the entire city's credit system might fail. Indeed, fire insurance was an absolute requisite "without which nineteenth century commerce was impossible," yet "commercial calamity" would surely follow unless something were done. Perhaps, it was thought, the London insurance companies would lower rates if they were fully apprised of the grandiose plans for the immediate future of a possible special water reservoir for fire emergencies and the now widespread use of the Racine in the city.[92]

A meeting at the Port of Spain Chamber of Commerce office in April 1899 intensified the malaise among the city's business leaders. The discussion was intended to deal with the escalating fire insurance rates but ended up pitting different groups of city merchants against one another on the basis of store location. A recent fire in Henry Street led to general recriminations against the city fire brigade, and everyone agreed that publicizing this lament was generally harmful. The conversation then shifted to the abnormally high frequency of arson in one section of Frederick Street, leading one speaker to express hope "that the insurance on buildings in that street would be doubled and in other parts of the town reduced." Were there "habitual criminals of a very bad type" preying on a particular group of stores, and what would the insurance companies conclude about rates if they were apprised of this hypothesis? Above all, with meetings like this one being held at the Chamber of Commerce, a body intended to promote business, it was "open to the world to ask, what kind of a commercial community is . . . Port of Spain?"[93]

Inevitably, the problem of the city's fire insurance rates boiled down to questions of water and water supplies. Early in 1902, in response to queries and pleading from various local agents, the British insurance companies (who claimed that fire premiums had been raised in a number of cities elsewhere to recoup losses for the 1895 Port of Spain fire) said that they were ready "to reduce . . . rates as soon as a proper supply of water" was available in the city "along with the erection of necessary reservoirs."[94] Then in May the city's insurance agents, acting under instructions from their head offices, raised fire insurance rates by a further 25 percent. This alarming increase led to speculation as to how far American and European manufacturers, in the light of Trinidad's prohibitive fire insurance rates, now would extend credit to Port of Spain merchants. It was even possible that suppliers might consider the city too dangerous a place to send their goods at all. Under these circumstances, blame had to be fixed and local culprits identified; it was futile to blame the insurance companies for making seemingly objective decisions about dangerous local conditions and taking cautionary, businesslike steps to account for them. So who was "responsible for having succeeded in placing the city in so unenviable a position"? The problem, a majority in the city agreed, was the government's inadequate handling of the water supply for Port of Spain. Despite promises and project announcements since the 1895 fire, the water supply had simply failed to keep up with a growing population, and now, among many other problems, the city's immediate economic future was in peril. Even the *Port of Spain Gazette,* the city's most conservative newspaper, in May 1902 called for radical action: "The population should in a united body protest against this ruinous lack of the necessary water supply."[95] The writers of these editorial remarks could not have known at the time that their advice soon would be acted upon in an extremely forceful manner.

Port of Spain's Water Supply

Trinidad's growing and increasingly concentrated human population at the turn of the century placed heavy demands on the colonial government for its piped water supplies. In 1900 water for the western part of Port of Spain descended from the original Maraval River reservoir, located in the uplands about four miles northeast of the city, in two cast-iron water mains. The eastern and smaller part of the city was similarly served by the St. Ann's River reservoir. In 1895 water engineer Osbert Chadwick had recommended tapping the potential water supply from the Diego Martin valley northwest of Port of Spain in order to augment the city water supply, and that plan eventually became reality in mid-1903.[96] San Fernando, Trinidad's second city, to the south, first received piped water in 1884, but its water supply remained unreliable; many in San Fernando continued to use wells and cisterns, and residents there complained on occasion of the sulfurous and oily tastes in their water. By 1900 major wells had been dug for some of the growing population centers running east of Port of Spain, and plans were under way for serving Tunapuna and St. Joseph with piped water.[97] But in the end, it was the Port of Spain water system, because of escalating fire insurance rates and unending complaints from all segments of the local populace, as well as for far more important reasons, that would merit local, regional, and even immediate Colonial Office attention at the turn of the century.

By 1901 Port of Spain's public water and sewerage system was beginning to show the overall signs of the wear and deterioration of a public service network that had been in place for a half-century. In February of that year some of the earliest laid sewerage pipes were being replaced with "modern galvanised iron pipes," and it was a wonder to some that the city's patchwork sewerage system had never been the source of a major disease outbreak.[98] Only the northern part of the city had sewers, and these were of "a not very modern type." In the rest of the city and the outlying areas, the contents of cesspits were periodically removed, then "more or less completely destroyed by burning, after which it [was] used as a fertilizer." A daily system of scavenging, or refuse collection, as in the region's other towns and cities, involved the carting of unwanted material away from houses and public buildings, and the city's gutters were swept and flushed daily, an operation that obviously called for a copious volume of water but that always fell short of expectations during the dry season in the first months of every year.[99]

Port of Spain's seasonal water shortages, as well as concerns about the overall quality of the city's water, naturally led to the periodic scrutiny of the character and cleanliness of the reservoirs above the city. To allow water to accumulate in the reservoirs in the dry seasons, authorities typically locked off the water mains serving the city for hours at a time, often during the nighttime hours. Bush fires damaged the landscape surrounding the St. Ann's reservoir in April 1891, further reducing the water for eastern Port of Spain.[100] Whereas government officials had no control over the rainfall (and little control over the bush fires that accompanied the dry periods), they were held responsible by a variety of critics for the reservoirs' cleanliness. The "filthy" and "putrid" character of the Maraval reservoir was openly condemned in mid-1889.[101] In the next decade government officials monitored villagers who resided along the upper reaches of the Maraval River that fed the reservoir; the rural peoples used the stream for bathing and laundry and even resorted to the riverbanks "for the purposes of nature" rather than using cesspits. These observations, in turn, led to dire warnings that the city's water supply itself could lead to "contamination of a deadly character."[102] It did not help that, even at the turn of the century, Port of Spain's drinking water still was not filtered through sand beds. In that regard it lagged behind even neighboring St. Vincent and Grenada, places that Trinidadians habitually looked down on in nearly every other way.[103]

Filter beds aside, the problems of undependable water quantities for the city, especially in the dry seasons, inspired public outcries against the government to provide more water for Port of Spain's growing human population. By the late 1880s words like "crisis" and "collapse" appeared regularly in official speeches and newspaper headlines, and private conversations must have been similar in terminology and tone. Public discussions, following reports by government engineers or visiting advisors, routinely dealt with the various projected quantities of water necessary to serve the city, emphasizing how many hundreds of thousands or even millions of gallons would be necessary for various purposes, both now and in the future. Without these totals, city residents would likely "be compelled to soon revert to the primitive methods of sewage-tainted wells and rain water tanks for our water supply." If a serious fire were to break out, it would most likely lead to "the speechless indignation that agitates the irate bystander at a fire, when the open hydrant refuses to play and the fire fiend laughs at the impunity conceded him to spread unnecessary havoc."[104]

So the pivotal water shortfall during the March 1895 Port of Spain fire, which was unexpected and unprecedented in its specific details, extent, and damage, represented, on the other hand, a predictable culmination of the problems and alleged governmental mismanagement that water supply critics had been pointing out for years. Five months after the fire, on August 13, a well-publicized public meeting in Port of Spain dealt resoundingly and negatively with the main provisos of the latest proposed water legislation that the government had developed in the wake of the fire. In addressing the main provisions of the pending "Port-of-Spain Water Supply Regulation Ordinance, 1896," the meeting's organizer, Robert Wilson, led those assembled to reject the proposed plan to install water meters in selected residences and buildings in the city (so as to measure and charge for water usage by volume) and to demand a far more voluminous water supply on behalf of the residents of the city.[105]

Wilson was an Englishman who was described candidly by one Colonial Office correspondent as "a man of broad views and radical sympathies who occasionally spends a year or two in Trinidad to look after his business." In the August 1895 meeting Wilson criticized relentlessly the government's proposal for installing water meters. The water meter idea had been proposed publicly three years earlier by a retired official of the Trinidad Public Works Department, and his suggestion had been countered with widespread scorn and hostility since then. Now, with the 1896 water act, the government again suggested that water meters be installed, but selectively and only in the houses of the well-to-do in the city, so as to discourage wasteful practices by the rich and thereby leave enough water for the poor. In a written memorandum published three weeks after the August public meeting, Wilson ridiculed the government's proposed water meter system. He claimed that there were no water meters "in any English town of any importance." Further, the proposal for a class-based system of installing water meters represented a "mistaken sense of economy" and could set in motion a transparent and pandering "Poor-Man-Protection" theory that would backfire in practice and which was a clumsy attempt by the government to foster interclass antagonisms to cover up its own blundering and mismanagement of the water problem.[106]

It was the problem of supply, or overall quantity, of city water that government planners and engineers had proposed to solve by installing meters. Water shortages could be relieved, they reasoned, not by providing more and more water—an increasingly difficult task given the finite physical resources of the island—but by making better use of the water already available. Thus, "waste" became one of the government bywords in describing the city's water supply issue. Exaggerated examples of water waste, each certainly containing a grain of truth, described the squandering of precious amounts of the fluid that, with proper conservation methods, could serve the entire community for its drinking, bathing, and fire protection needs. Rich people splashed about in their oversized "plunge baths," taps were left running in barracks yards all day creating inaccessible quagmires and sodden earth, and water ran endlessly in the city drains to the sea when it could have been preserved for more practical purposes.

Political scientist Alvin Magid has characterized the ineptitude of the Trinidad government in handling the colony's water situation late in the nineteenth century: "Alternately permissive and stern in its administration of the Port-of-Spain waterworks before 1895, the Crown Colony government brought contempt on itself."[107] Further, he has shown how the seemingly innocent, even trivial, issue of proposed water meter installation in Port of Spain in the next several years was a crucial material element interwoven into a many-stranded fabric of public discontent and eventual explosive outburst. In any event, the 1895 protest organized by Robert Wilson against the government's mishandling of the local water situation appears to have nurtured a relentlessly negative public attitude, reinforced in everyone's mind whenever a fire—however small—broke out in the city. This antigovernment mind-set was further adopted as a cause by various disgruntled and otherwise unaffiliated interest groups and was helped along by the dithering of a maladroit government administration. The concerned and otherwise diverse groups opposing government water proposals included middle-class white planters, Spanish speakers, and the Ratepayers' Association, a group of property owners formed in 1901 who acquired considerable political muscle early in the following year "when Emmanuel Lazare, a radical black solicitor, joined its ranks."[108]

A major fire broke out in Port of Spain late at night on March 26, 1902, in the lumberyard of the Trinidad Shipping and Trading Company, eventually destroying the entire block from Richmond Street to the waterfront. While it seems futile to identify a single occurrence as the unequivocal cause that set a succession of events in motion, this fire appears particularly important in helping to explain what followed during the next year. Further, it matters less that the lumberyard fire was judged the work of an incendiary than that the lack of available water allowed the fire to "have its own way" until much too late. As usual in the dry season, city water was to be cut off for a few hours in the middle of the night of March 26, and placards around Port of Spain announced that water would be available until midnight. But the water actually had been stopped earlier in the evening, which was apparently also usual, reportedly by the already unpopular director of the Public Works Department. So the fire brigade, whose members arrived promptly at the scene of the fire shortly after eleven o'clock, stood by helplessly for roughly a half-hour trying to coax water from the hydrants, but they "could not get one drop to cope with the roaring flames."[109]

The lumberyard fire, combined with a smaller blaze that destroyed a residence on Charlotte Street a few days later, stoked the editorial embers at the usually conservative *Port of Spain Gazette*, which told the world about "the water scandal" that had led "to the alarm and irritation among the public at the insufficiency of the water supply of the city and the manifest incapacity of the government in a quarter of a century to achieve anything serious in the way of improvement."[110] This rhetorical flourish, echoed by the other Trinidad papers, effectively extended the Port of Spain water issue to the British Caribbean as a whole. Especially in the nearby small southern islands, many of whose citizens recently had emigrated to Trinidad in increasing numbers, the Trinidad water problems were watched with great interest. A Grenada newspaper went so far as to join the anti-government-of-Trinidad chorus, lashing out against the "huge swindle" being perpetrated against the people of Port of Spain and asserting that, had taxpayers' money been spent wisely over the years by the profligate government bureaucrats in Trinidad, "hundreds of millions of gallons daily" of fresh water now would be pouring into the city for use by its citizens.[111]

At a mass public meeting in Port of Spain in June 1902 called by the Ratepayers' Association, Emmanuel Lazare lashed out at the government for its water policies as well as other shortcomings. Four months later another such meeting focused on the drawbacks of yet another newly proposed government water ordinance. Intending to fold several old ordinances into a new policy, the government plan centered, as far as those addressing the crowds were concerned, on the proposed—and ever unpopular—idea that water would be conserved if meters were installed. At the meeting one spokesman after another denounced the plan, pointing out, among other things, that even if water meters were affixed to the pipes entering the houses and other buildings of the rich, the extra charges would be paid in the form of additional rent charged to everyone else. The rhetoric at the latter meeting also took on a broader tone, with sweeping generalizations about colonial oppression. Speakers suggested that if tax rates like those in Port of Spain were imposed on English citizens, they would yield direct benefits in public services rather than the poor physical amenities accruing to residents of Trinidad's capital city.[112]

In October 1902 an epidemic of what was perhaps a form of smallpox (which local health authorities insisted was chickenpox to avoid quarantine restrictions) began in Trinidad, and in the following months the progress of the disease, combined with restricted water supplies, probably went far in embittering the city's working peoples and turning them against the government. By the following February 387 cases of this eruptive fever had been reported, and probably many more were unreported. As the dry season commenced, moreover, the lack of water in city drains was said to have intensified the spread of the disease: "The drains were unswept, and the dry season wind was sweeping up the dust and the germs of the contagion contained in the crust of the drying sores of people recovering from the eruption and distributing them broadcast." Then in mid-February 1903 government water restrictions (an attempt to curb use during the dry season) were loudly denounced as contributing directly to the further spread of the disease; moreover, the restrictive act was colorfully described as the onset of a "reign of terror" in an inflammatory memorandum sent by the Port of Spain Ratepayers' Association to London.[113]

The men leading the Trinidad government, under pressure from London to curb the accelerating protests over the water issue in the city as well as absorbing daily criticism from local sources, came up with the Port of Spain Water Ordinance early in 1903. The bill had its first reading in the local legislative council on February 23. Water meters would indeed be part of the new plan, but individuals with grievances could seek redress; the latter point soothed nobody among the local citizenry. A mass meeting, again called by

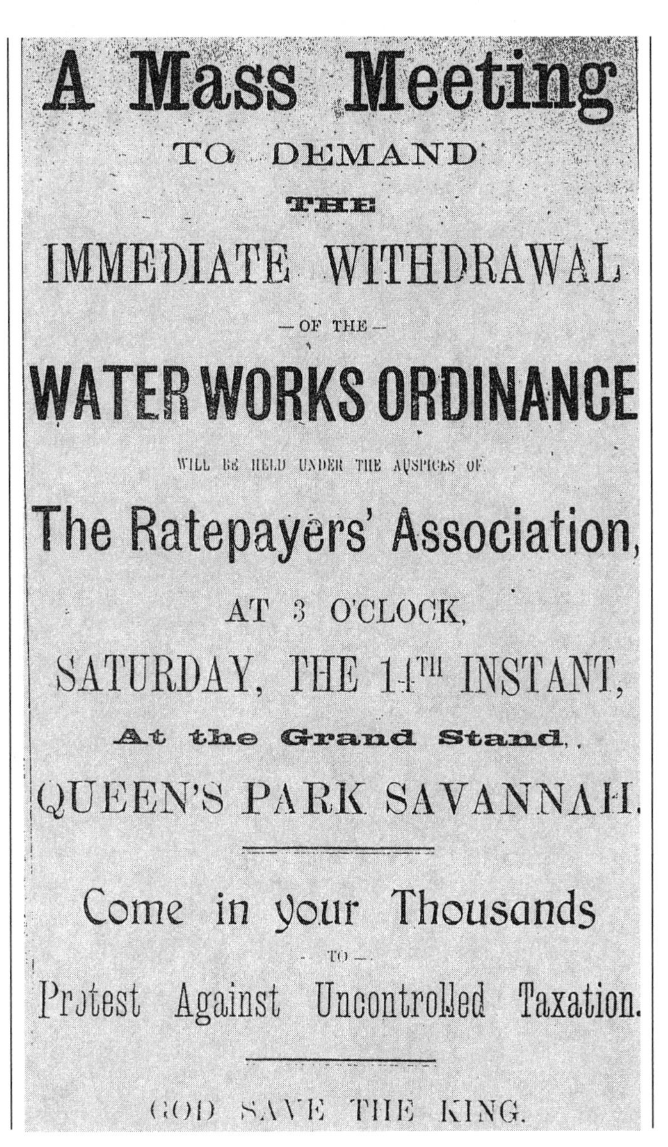
A Mass Meeting
TO DEMAND
THE
IMMEDIATE WITHDRAWAL
— OF THE —
WATER WORKS ORDINANCE
WILL BE HELD UNDER THE AUSPICES OF
The Ratepayers' Association,
AT 3 O'CLOCK,
SATURDAY, THE 14TH INSTANT,
At the Grand Stand,
QUEEN'S PARK SAVANNAH.

Come in your Thousands
— TO —
Protest Against Uncontrolled Taxation.

GOD SAVE THE KING.

Newspaper advertisement protesting the government water ordinance (from *Port of Spain Gazette*, March 13, 1903)

the Ratepayers' Association, at the Queen's Park Savannah on March 14 addressed the new water bill head-on. The published announcements for the meeting urged citizenry to "come in your thousands" and ended with "God Save the King," thereby patriotically suggesting that protesting what the government was doing was really the responsibility of all good citizens. The meeting's speeches were the most inflammatory yet, pointing out that government henchmen, in order to implement this new water law, would invade private residences, thereby violating the sanctity of individual dwellings, even interfering with "your wives and daughters" in "their baths."[114] Yet this bombast was no worse than public pronouncements two weeks earlier suggesting that government "vandals" who were the "respectors of no one" would, if these new plans were adopted, invade the yards and houses of every private citizen and shut off running water.[115]

One week after the mass protest at the Queen's Park Savannah, the Red House in Port of Spain was the scene of the March 1903 "water riot." News of the protests shook the region, caused police from elsewhere as well as British bluejackets to be deployed, and drew the attention and ire of the London Colonial Office. As for any such event, several interpretations could be forwarded to account for the disturbances—explanations based on oppression, race, class, and so forth. In many ways it was also an outburst of pent-up feelings against what was interpreted on the ground as the denial of a fundamental resource: "The people of Port-of-Spain . . . from 1850 to 1875 enjoyed . . . an excessive supply of water. . . . The result was that they looked upon such a supply as their right."[116]

The provision of piped water had allowed the population of Port of Spain to grow to an estimated 50,000 in just a few decades, attracting those from the island's countryside as well as immigrants from nearby islands, so that the city's water supply had become a regional issue. The agency controlling the volume, quality, and infrastructure for the city's water, Trinidad's colonial government, furthermore, was responsible for the maintenance and extension of the water supply that sustained the city's populace. Water was therefore not simply an amenity, and it was the government's acknowledged duty to safeguard and continuously improve its supply. The government's apparent violation or lack of protection of such a fundamental right was, further, a call to action. Despite the long-simmering disgust of the city's citizenry, the action itself required a catalyst or spark, and doubtless the most memorable feature of the resulting riot was the burning of the Red House in the center of the city.

The government's denial of the fundamental right of preserving and extending the people's water in Trinidad in March 1903 and the resulting trouble could be interpreted as a conflict over resource management. Formerly supplied by nature, water now was under the maladroit and, according to many, capricious and even contemptuous supervision of the local government. Lacking control over water, the people of Port of Spain responded by protesting with fire, creating control problems for the government of an altogether different magnitude.

7

Fires of Protest

DETAILED DESCRIPTIONS OF THE EXTENT, damage, and probable origins of a huge and destructive fire at Colón in the Colombian province of Panama in late September 1890 were published in several official gazettes and newspapers of the British Lesser Antilles. These articles worried property owners throughout the Caribbean because, among other regional parallels, only three months earlier much of Fort-de-France, Martinique, had been similarly destroyed by fire. The Colón fire began a few minutes after midnight on September 23, 1890. It was fanned by a stiff breeze and initially fueled by the combustion of several large vats of alcohol and kerosene in one storage building; the fire consequently expanded in size, gained temperature and strength, and eventually destroyed several blocks of warehouses, the city's market area, much of Colón's commercial center, and an estimated 2,000 houses. A thirty-minute rain shower about 3:00 A.M. subdued the flames momentarily. Although the fire did not destroy the city's ship docks, several steamers moored at Colón fled the port about 1:30 A.M. and did not return until after dawn when the fire had died down. At 8:30 the next morning a special train arrived from Panama City with the governor and a detachment of troops to maintain order. The initial estimates of the overall fire damage ranged between $2 million and $3 million.[1]

Beyond the distress created by losses in property and housing was the near-certainty that incendiarism was behind the Colón fire, and this led to the related concern that it could easily happen elsewhere. Further, although the perpetrators were neither identified nor prosecuted, everyone agreed that the fire was a clear case of arson resulting directly from social protest. In the days before the fire, local railroad wages in Panama were reduced from $1.50 to $1.00 per day, a move that inspired unorganized but intense public disorder. Angry, shouting men armed with sticks and knives, especially immigrant "Jamaican negroes" working in the city, roamed the streets in small groups. Small knots of these men, "maddened by rum and the intoxicating effects of a few days march to the beat of the Tom Tom," then coalesced into larger crowds. Finally, as described in the *Antigua Standard* a few days later, the destructive capacity of this large group of angered men, a potential for mayhem that had been "slumbering" within the recesses of the maddened crowds, suddenly "burst its bonds," leading to arson and the disastrous fire at Colón.[2]

The phrase "burst its bonds" in the Antigua newspaper description suggested that there existed a social, not only a chemical, combustion point, which helped explain how the Colón protest suddenly had flared beyond control. In cases such as this one, fires and riotous behavior always seemed to go together and to draw strength from each other. Nor was it particularly important which half was the cause and which the effect. Perhaps at a crucial instant the sudden presence of fire had raised the Colón crowd's protest activities to an uncontrollable level. Conversely, perhaps there had been a moment when fire had been sparked directly from the angry crowd's riots and demonstrations.

One can perhaps make too much of the wording in a single newspaper article by casting it as an interpretive expression of the interrelationships of West Indian fire and riot at the turn of the century. Yet two similarly vivid published descriptions of fires as flashpoints of local riots similarly reinforce the idea that group grievances or protests seemed to possess kindling points susceptible to being ignited by fire. In February 1896 a Barbados newspaper described the origins of the momentous St. Kitts labor riots that "like a flash of electricity" spread across the small sugarcane island, leading to blazing cane fields amidst social tumult.[3] Similarly, the 1903 water riot in Port of Spain, which resulted from many of the background events discussed in Chapter 6, was not necessarily inevitable, nor would it have taken the same devastating direction without fire. The *Voice of St. Lucia* stated that the dire situation in Trinidad in late March 1903 had become intolerable for Trinidad's people and that the massive organized protest at the Red House in Port of Spain had been "blown into flame," leading to the rioting and killing that shook the entire region.[4]

Yet an excursion into semantics was hardly necessary to remind local Caribbean officials that combining angry crowds with fire nearly always meant trouble. The contested *cannes brulées* processions of course preoccupied Trinidadian officials every February. The disturbances such as those on Guy Fawkes night in Grenada in 1885 and the St. Vincent protests late in 1891 were reminders that fire could lead to riots, and vice versa. But especially in the economically depressed atmosphere of the eastern Caribbean at the end of the nineteenth century, questions as to the immediate material or emotional circumstances that moved crowds of protesters from angry groups of malcontents to uncontrolled rioters were eminently practical, useful, and relevant. Starvation wages (or the loss of wages altogether) on a particular island could lead to misery, discontent, and grumbling among crowds whose activities—sparked by the wrong conditions—then could get out of hand. Barbados chief justice Conrad Reeves, for example, sensed a drift toward "crisis" on his home island. Accordingly, he proposed in 1886 that the colony's treasury underwrite local plantation loans so that laborers' wages were paid. London officials considered Reeves's idea sound, and they spelled out for one another what this crisis on Barbados could lead to: "We may admit that there is a danger of a good many estates going out of cultivation; and when that happens in Barbados, the too large working population, which in the best of times does not get continuous employment even at a very low wage, is reduced to starvation—incendiary fires become numerous, and there is a rising which the troops can with difficulty put down."[5]

"Risings," whether caused by fire or other immediate reasons, had ample precedent throughout the Caribbean. It seems certain that British colonial officials of the region routinely lost sleep over the possibilities that their particular island, given the wrong set of events, might literally go up in flames. The revolutionary war in Cuba in the late 1890s—its pitched battles, scorched earth campaigns, and eventual expulsion of the Spanish all duly reported in British West Indian newspapers—demonstrated what could happen in even the largest places in the region. Certainly the British navy would intervene if one of its colonies were similarly threatened, but what might happen before the naval destroyers and bluejackets arrived or, more likely, after they left? Perhaps most important, these possibilities were not without recent precedent. The Morant Bay rebellion in Jamaica in 1865 and the Confederation Riots in Barbados in 1876 had attracted regional attention and alerted the London Colonial Office that islandwide risings were indeed far from impossible.[6] The infectious character of and mimicry encouraged by fires set openly in the heat of passion just might ignite such events anywhere, even in those few islands where the sugar depression was not so severe. In mid-1882, for instance, when anonymously produced placards went up around Roseau, Dominica, deploring the increase in local road taxes and demanding that the town's shopkeepers import plenty of "kerosine and matches in large quantities," frightened local officials took special notice.[7]

Was it the presence of fire or was it some other cause, or perhaps a combination of causes, that transformed crowds of disgruntled men into destructive mobs? This was far from a merely rhetorical question in late Victorian Britain and the rest of industrial Europe at the end of the 1800s. Dozens of books and pamphlets authored by academics, journalists, clergy, and even novelists addressed this issue. A scientific understanding of group behavior and related questions, furthermore, was considered not only possible but also desirable because it was the first step toward the control of crowds and the elimination of the problems destructive mobs could create. Perhaps the best-known "mob" author was Gustave Le Bon, a French sociologist whose study *The Crowd* was published in 1895, with the first English translation appearing the following year. Le Bon's work and those like it were widely known in literate circles on the Continent that included the men of the London Colonial Office, whose reading and thinking influenced their management and control of Britain's possessions in the Caribbean and elsewhere.

Le Bon's prose described crowds in as frightening a manner as possible and depicted potentially destructive social entities possessed of collective minds. According to Le Bon, individuals succumbed to the moods and actions of the group, which could be swayed by immediate and "contagious" events: "A crowd is at the mercy of all external exciting cases. . . . It is the slave of the impulses which it receives."[8] Whereas Le Bon did not specify fire as one of these impulses, everyone knew incendiarism as a common catalyst that animated and energized Western European social disturbances in the late eighteenth and early nineteenth centuries. Rural arson was a particularly common weapon in English agrarian uprisings in the early 1800s, and in the middle of that century an Italian writer noted that the "dreadful calamity of fire" was "a peculiarly British device," a well-established means by which the weak could cope with, or at least react to, the demands made upon them by the strong.[9]

The fin de siècle writings of Le Bon and other social theorists dealing with crowds and social control held an immediate and timely fascination among literate Europeans of all social classes. Late in the nineteenth century, large and visible groups and crowds of working peoples were perhaps the social hallmark of an ever more industrial and urbanizing Europe. The very presence of these groups of laborers, further, exerted considerable social and political pressures on capitalists and politicians. As contemporary social thinkers and newspaper columnists described the growth and potential menace of working-class crowds, literate Europeans, as individuals and as members of reading societies, were reminded of the multivolume study by Hippolyte-Adolphe Taine, *Origins of Contemporary France*, as well as Thomas Carlyle's *The French Revolution*. Both studies reinforced and emphasized the powerful and genuinely transformational role played by the mob during the French Revolution, and one century later parallels between the social and economic inequities of prerevolutionary France and the contemporary post-industrial societies elsewhere on the Continent seemed to abound.[10]

In the Caribbean, crowds or groups of working peoples were hardly new phenomena; enslaved Africans and, later, indentured Indians and others had been imported to the region, not to act as individual yeoman farmers or other kinds of independent decision makers but to take their places as atomized members of labor gangs. But these plantation "crowds" always had presumed the kinds of top-down discipline and coercive direction that were the socioeconomic sinews of plantation organization and social control, Caribbean customs that had hardened into legal sanctions and that protected officials and planters from the trouble that undisciplined crowds could create. In the decades immediately following emancipation, the diffuse spatial character of rural populations housed in villages adjacent to scattered sugar estates on individual plantation islands usually precluded the coalescence of large unsupervised crowds. Late in the nineteenth century, however, as depression conditions pushed more and more rural West Indians via improved road networks into capital towns and cities, the potential for and presence of potentially troublesome urban crowds were increasing.

The small scale of individual Caribbean islands also could easily facilitate the spread of trouble, a fact of Caribbean geography lost on no one, especially after the rebellion in Morant Bay in Jamaica in 1865. Although small island interiors were often rugged and impenetrable, better transportation facilities were opening them up. How long would it take for determined crowds bent on trouble to take over an entire island colony of only 100 square miles, even if the interior were mountainous? Insular officials were therefore wary of the movements of people, especially sizable crowds, in rural areas. Possible coalitions between rural dwellers and residents of the cities and towns, for economic reasons or celebrations or other special events, were noted with more than passing interest. Communication or signals back and forth among individuals and groups, other than word of mouth, were, however, very limited. Perhaps the most common such signal was the use of fire, which could convey a range of emotions, depending on the local circumstances, and which was visible, especially at night, from one part of an island to another and in some cases from one island to another.

Of course, Caribbean crowds were not composed of the proletarian coal miners, millworkers, or shipyard laborers of Western Europe, the stereotypical "laboring men" about whom Le Bon and, later, Eric Hobsbawm and others would write.[11] But ever larger crowds of West Indian men and women, often unshod but wearing floppy hats and cotton shirts, trousers, and dresses, now were commonplace in the port towns and cities of the eastern Caribbean. Both periodic and random events on every island provided innumerable opportunities for crowds to form. Saturday morning markets, ship arrivals or departures, special events such as the Corpus Christi parades of St. Lucia and the queen's birthday everywhere, or the possibility for wage labor invariably swelled crowds, usually to the dismay and unease of insular officials. In June 1888 the arrival by ship in St. Vincent of the deposed and exiled West African king JaJa of Opobo was planned for a Sunday morning so as to avoid drawing too large a crowd at the Kingstown waterfront; but the news was leaked, and a "crowd of considerable dimension" nonetheless appeared.[12] Two months later, before the celebration of the fiftieth anniversary of emancipation in Castries, St. Lucia, crowd-fearing officials hoped that celebrants would exercise a "self-restraint and a self-respect that will silence forever the detractors of the vast majority of the inhabitants of this island."[13] In 1895–96 during the repairs of the pierhead in Bridgetown, Barbados, periodic daily wage work for an estimated 200 men regularly drew crowds of up to 500, and "on one occasion a guard of . . . forty police, including horsemen, had to be posted in order to prevent the rush."[14]

When genuinely large crowds gathered for even obliquely social or political reasons, officials feared the possibility of unrest. Late in 1889, when Sir John Gorrie, the reformist chief justice of Trinidad and Tobago with a wide following among black and Indian workers, landed at the Port of Spain waterfront, a crowd estimated at 10,000 turned out to greet him. Bands played, flowers were strewn in his path, and "a hundred hands" unharnessed the horses to his coach so that it could be pulled through the city streets by adoring throngs.[15] Yet the happiness and admiration displayed for the few beloved officials in the region in the tense social atmosphere of the late 1800s could easily turn ugly when large crowds were involved, and especially when fire was at hand to ignite group passions and turn them in destructive directions. Earlier in 1889, when Gorrie visited Scarborough, Tobago, similar group revelry greeted his arrival, but unlike the peaceful celebration that would mark his arrival months later in Trinidad, the Tobago visit was punctuated by several fires set openly on the estate of a particularly disliked planter. "The idea it leaves is that there has been some great and ongoing oppression of the people, and that a great redeemer has come among them riding in a fiery chariot."[16]

At the end of the nineteenth century, the parallels between European recollections of the French Revolution and the overarching pan-Caribbean fears of another Haiti, perhaps in a nearby colony or even on one's own island, could not have been more obvious. Late in 1899 the *Times* of London ran "The West Indies," a series of newspaper articles describing the economic plight of the area. All of the region's well-known problems—such as the bounty depression, the antiquated sugar production equipment, the competition with newer and larger sugar colonies in the tropics, and the spent insular soils—were reiterated in the articles for interested British readers. Yet the final article in the series pointed out what was really at stake if the impoverished and neglected British Caribbean colonies continued their downward slide: "Behind the economic question stands the social question, behind the economic ruin of Barbados and Antigua stands the ghastly political spectre of Hayti—a spectre which eternally haunts and terrifies every one of the West Indian colonies." The article concluded by describing this "spectre" precisely; it was "the terrible allegations which depict Hayti as a black Inferno."[17]

Learned British readers and Colonial Office officials already were well aware of Haiti's possible precedent for the depression-racked islands of the British Caribbean. Frightening reports abounded about what had happened there in the 1790s, such as those written by James Anthony Froude, an Oxford history professor who traveled through the British Caribbean in the mid-1880s and then wrote an interpretive account of his travels titled *The English in the West Indies*. His racist (even in the context of the late 1800s) descriptions of the people of the British islands were compared in nearly every case to Hayti, a place where almost yesterday—according to Froude's lurid descriptions—black mobs had chased down white women in order to impale their babies on spears.[18] These horrifying images concocted by Froude and others were never complete without the obligatory backdrop of flames and fire, a medium that—once out of control—could act as a catalyst not only to kindle resentment and peaceful protest into riot and destruction but which, in the obvious case of Haiti and perhaps now in the small British islands, could lead, infernolike, to the savage destruction of an entire island colony.

Lengthy excerpts from Froude's book appeared in several British Caribbean newspapers of the time, and the accompanying editorial commentary about his scholarship was not always favorable. Yet innumerable other references in British West Indian papers, depicting what things were like in Haiti now and in the past, never let readers forget what might happen at home if local unrest escalated into fiery riots. An exemplary article in a Dominica newspaper in August 1888 described recent outbreaks of fire in Port-au-Prince and the explosive political atmosphere in Haiti as an almost cause-and-effect relationship: "The revolution, which everybody expected would have followed the late disastrous fires at Hayti, has broken out."[19]

Was the malicious use of Caribbean fire as a revolutionary instrument by uncontrolled mobs in Haitian history, and now threatened elsewhere in the Caribbean, a peculiarly French perversion? Selective newspaper reading might lead to such a conclusion. In May 1891 the editors of a Dominica newspaper published a translated article from *Le Propagateur* in neighboring Martinique that reported the posting of placards by urban dissidents around the capital city with the ominous suggestion, "Bruler Saint Pierre." Such a blatantly public exhortation should not, according to the Martinique paper, create undue panic, but rather, "precautions should be redoubled."[20] According to a report in the *Voice of St. Lucia* in early 1900, rural rioters demanding higher wages on the sugarcane plantations of Martinique attacked and burned local factories; the army was called in, nine persons were killed, and several were wounded.[21] One can only imagine the suggestive rumors and information passed by word of mouth about the threatened burning of St. Pierre and the agrarian riots among the black, working-class friends and relatives in Dominica and St. Lucia and elsewhere whose brisk and formally illegal traffic back and forth with Martinique was daily carried on via a flotilla of small sailing vessels.

Although frightening and suggestive, French and French Caribbean influences paled in comparison with how fire and riots might ignite, and in many ways lay bare, the "African" passions and wickedness that lay just below the thin veneer of European influence throughout the region. In the same way that unsupervised village cultivators in forested Caribbean highlands were seen to revert to primal African ecological barbarities and thereby ruin insular environments, all of the region's descendants of Africans—if many accounts were to be believed—could, given the wrong conditions, return to barbaric forms of behavior almost overnight. This primitive condition, significantly, was often illustrated by the newspaper accounts of the thoughtless, unbridled, and savage ways in which common people behaved in the presence of fires, so reminiscent of despicable African practices.

In the two years preceding the 1885 Guy Fawkes riots in St. George's, Grenada, the November bonfires there had far less to do, according to some, with commemorating the Gunpowder Plot than with exhibiting primitive rites, liberated by fire, from the Dark Continent. In November 1883 the throwing of fireballs around the capital town's Market Square was accompanied by "the yellings and the antics" that "so much resemble the war-dances of the uncivilised African tribes." The following year the fire danger to the town reemerged, and "the way in which these people enjoy themselves savors very much the heathenish festivals and war dances of the savage tribes of Africa."[22] In 1887 in Trinidad, an ominous sense of dread followed the reports of a rise of African-based obeah and witchcraft practices in Port of Spain and closely related rude and "uncivil" acts among the black populace. Here were indications of impending trouble that could spread quickly through the whole island and a foreboding that perhaps could be expressed only by a fire metaphor: "The danger of disturbances on a large scale may not be so remote as some might think. That a dangerous train of combustible matter is spread all over the Island is manifest . . . from every quarter; the slightest spark might ignite the inflammable element."[23]

Allegation, commentary, and concern about transplanted African savagery, fire-illuminated and otherwise, had broad socioeconomic significance in the British Caribbean at the turn of the twentieth century. Preoccupied with the larger and more important areas of the empire, the London Colonial Office entertained the possibility for varying degrees of local political and economic autonomy in the Caribbean colonies. Although possible changes never included the giving over of island economies and governments to their black majorities, even the relaxation of land policies threatened the reigning plantocracy. So warnings that greater local control might unleash chaotic African behavior that would in turn send island economies into downward spirals were not uncommon. When the 1897 royal commissioners arrived in Dominica to conduct hearings about economic conditions on the island in light of the sugar depression, some of the mixed-blood locals complained that they would be prejudged because earlier commission testimony in London already had painted a "lurid picture of a black and coloured West India given over to riot, anarchy, and bloodshed as the inevitable sequel to a failure of the Sugar industry."[24] The mental imagery associated with these lurid pictures doubtlessly would have been incomplete without a background of flames.

Although ostensibly African-based behavior was condemned throughout the region, British officials in the southern Caribbean, where considerable numbers of indentured and post-indenture Indians resided, identified similarly objectionable, even barbarous, cultural activities that had been transported from the subcontinent. In central Trinidad, where open land and forest came together, agricultural officers decried Indian villagers' land clearance campaigns and preplanting burning activities because they were said to lead to soil infertility.[25] Early in the 1880s Indian "fire-walking" ceremonies over pits of hot coals at Peru Village were described as a practice brought from South India. Whatever its origin, it was roundly condemned in the local press as "degrading" and "vile," even "workings of the devil." Less fearsome, although still somewhat mysterious, was the practice of Divali, the festival of lights, in which villagers and townspeople alike illuminated their house and shop windows with tiny flickering flames produced by lighted cotton wicks inside earthen vessels filled with coconut oil.[26]

Whereas occasional commentary about Indians' village-level cultural practices was hardly grounds for widespread concern, the progressively violent behavior by Indian estate workers was. Work stoppages and strikes on rural sugarcane estates by indentured laborers in Trinidad in the 1880s and 1890s were marked by "burning the canes or mistreating the draught animals" as acts of resistance.[27] Reports of these events were particularly distressing for London. A leading consideration by the Colonial Office in the depression years was the welfare of the region's transplanted Indian population and, by extension, the nature of the ongoing indenture system itself, for which London was directly responsible. Of particular concern along these lines was the testimony by Supreme Court Judge Nathaniel Nathan to the visiting royal commissioners in 1897 when they came to Trinidad. Nathan predicted that a collapse of the local sugar industry would lead to "*organised violence* among the East Indian population thrown out of work."[28] Those in power obviously dreaded the spread of such violence from one estate to another, whether or not it involved Indians. Officials thereby attempted to curb the visible and auditory manifestations of any kind of outburst, protest, and riot—manifestations that of course involved the sight and smell of fire but also included noise.

In the late 1800s in the British Caribbean planters and colonial officials went to extraordinary lengths to control the sounds from angry public gatherings, especially noise that carried with it the potential to heighten emotions and swell crowds. One reference after another condemned group chanting, cadenced yelling, the clatter and scraping sounds made by sticks when beat against the wooden houses that fronted parade routes, or various forms of informal ("rough") music that attracted, enlarged, and intensified the emotions among members of ominous crowds. The sounds also were simply irritating, as in Castries in the Christmas season of 1905: "There can be no thought of interfering with the amusement of the masses, but the strain of one's nerves of one unvarying note on a shrill fife with a fast and furious drumming accompaniment is apt after a few hours to reach snapping point."[29]

Particularly eerie and threatening were the dolorous moaning sounds made by blowing on conch shells, signals that traveled long distances throughout rural areas of the region, often called people together, usually indicated trouble, and seemed to awaken the past. The disturbances in the windward La Plaine district of Dominica in April 1893 saw a contingent of British bluejackets from the destroyer *Mohawk*, augmented by a motley gathering of police constables from nearby islands, surrounded by a "mob" of angry, rock-throwing Dominicans protesting unfair government taxation policies and economic neglect. During the melee the marines, said to be fearing for their lives, shot four men dead at La Plaine village and wounded four women there, an event roundly condemned throughout the region. When the armed military and police force first arrived at La Plaine by boat from Roseau, the local crowd was said to have had materialized rapidly, summoned by the sound of conch shells being blown "from ridge to ridge."[30]

Insular officials considered drumming notably troublesome. The British Caribbean colonies maintained antidrumming laws very late into the nineteenth century; the statutes originated during slavery and subsequently were modified in the years thereafter. But these regulations were not legal anachronisms. Throughout the region authorities imposed monetary fines for the public "disturbances" resulting from village drumming.[31] Yet the real concern about drumming probably had less to do with public noise and disturbances than with drumming's emotive power to coalesce and energize. In part because it was outlawed, drumming seemed to carry with it an angry tone of resistance to authority, and it telegraphed and intensified warnings. More important, it resurrected passions that could take violent directions. Of course the African origins of British Caribbean drumming, in both rural and urban areas, were universally acknowledged and publicly discussed. Much of the archival evidence dealing with authorities' acknowledging these origins dwelt on drumming's evil and primitive character and suggested measures for noise relief or even the destruction of drums along with their offensive sounds. More positive descriptions of drumming included the one provided by Dominica's Governor Hesketh Bell in the late 1880s: "I once had as near neighbour an old African named Moses, whose particular forte was the tom-tom, and every night from dewy eve to early morn did my dusky friend wake the hours of the night with his doleful performance."[32]

"One Vast Wall of Fire and Flame": St. Kitts in February 1896

Written reports of the late nineteenth-century British Caribbean fires set in anger and protest describe flames and accompanying shouts and music and other sounds, multiple sensations that, taken together, must have been unforgettable. The culmination of the fiery labor riots in St. Kitts in February 1896 saw striking Basseterre dockworkers joined in the town by estate workers who already had been demonstrating for several days in rural areas of the island. According to Leopold Moore, the U.S. consular agent who resided in Basseterre, individuals within this surprising and unnerving coalition made several attempts to "fire the town," an effort thwarted only by the timely presence of bluejackets from a British warship. According to Moore, the surging crowds "behaved in a most disorderly manner all day continually blowing 'the shell' (a sort of horn made from a shell which means open defiance of the law) and acting in a riotous manner generally. They were joined from time to time during the day by gangs of laborers from the estates, marching in on the town from the country armed with sticks and stones to the tunes of their own native music."[33]

The St. Kitts troubles began in the first days of 1896 with the commencement of the sugarcane harvesting season. The effects of the depressed London sugar prices, leading to low wages in all of the islands, had been compounded on St. Kitts late in 1895 by the imposition of a burdensome local tax for water piped to dwellings in the country districts. General disgruntlement then led to strike actions and picketing on two sugarcane plantations east of the capital town. On the night of January 28, deliberately set sugarcane fires broke out on the two estates, attracting a "sizeable throng of onlookers." The crowd swelled, and the police summoned to maintain control soon found themselves facing a group of people twenty times their size armed with sticks and rocks. More ominously, a number of the protesters, and others who had joined them, were intoxicated not only by the consumption of rum but also by the presence of the "fire demon."[34]

Two weeks later, on the opening day of the 1896 cane harvest season, sugar mill workers on the same two estates refused to turn out for work. News of this work stoppage circulated quickly through the plantation lands and had "an electric impact on the estate workers of the island." Incendiarism then spread rapidly in the St. Kitts countryside. Noisy bands of striking men were said to control the roads at night, parading with drums and armed with heavy sticks and lighted torches. A frightening report of events on the night of February 12 appeared in the St. Kitts newspaper: "Quite a rebellion we had here last night. . . . For half a mile, nothing could the eye see but one vast wall of fire and flame. Along the public highway . . . a band of fifty roughs with sticks and stones, to the accompaniment of various instruments of music, the conchshell—with its hideous thunder—unmolested was parading. At intervals . . . came the crackle of burning cane and . . . the shouts of the maddened rabble."[35]

Urgent telegraph messages to Antigua from the officials and planters on St. Kitts led to the dispatch of the HMS *Cordelia* and its contingent of British bluejackets. The warship arrived at the Basseterre roadstead, just south of St. Kitts's small capital town, in the early morning hours of February 17. On the same day crowds of striking estate workers from the countryside, whose marches had heretofore been confined to rural areas, descended on Basseterre. The police attempted to block their entry into town, but the rural strikers eluded the policemen and joined forces with Basseterre dockworkers who quit heir jobs, demanding better pay. These rural and urban working peoples were then joined by petty laborers, market women, and the unemployed of the town, forming a swollen crowd that seemed to draw energy from the presence of the *Cordelia* and that "paraded the streets of the town unchallenged for the remainder of the day, armed with long sticks and cutlasses, and followed by a musical band playing fifes and drums and blowing on conchshells."[36]

As the growing crowd marched through the streets of Basseterre, the St. Kitts inspector of police sent a detachment of constables to the leeward side of the island to try to quell reported rioting and looting at Old Road Town. When they returned to the capital, the police ran into opposition and sustained injuries from stone-throwing demonstrators. Back in Basseterre, the marching crowds, now wielding torches, greeted the arrival of the local administrator's carriage with a shower of rocks, and as night fell, they began to smash street lamps and attack the shops owned by the "better citizens" of the town. Targeted rum shops and provision stores were set ablaze. Rioters entered several houses. When members of the fire brigade arrived, individuals in the crowd attempted to cut their fire hoses. The British marines did not, however, land until late in the day, about the same time the administrator's carriage was being stoned. Although he earlier had received orders to land his men, the commander of the *Cordelia*, Captain Bourke, refused to offload troops to help until he received an urgent personal request to do so.

The ensuing melee in Basseterre that night pitted "surging crowds" of black Kittitians against the British bluejackets, whose numbers were augmented by local police. The fight was backlit by flames from the buildings that had been torched. Throwing rocks and wielding sticks, the black crowd inflicted wounds on nearly every marine. Although the bluejackets were far better armed than the rioters, Captain Bourke, in his after-action report, noted the use of firearms against his troops. At several points during the struggle, the marines fired volleys into the air intended to ward off surges and attacks by the rioters. The violence finally began to subside after midnight, and marine sentries occupied strategic locations around Basseterre, attempting to establish order street by street. But the insurrection did not end until 3:00 A.M. By that time two rioters had been shot dead by the marines and five were wounded. Eleven stores were burned, and the town and its leading citizens were forever shaken.

At the end of the Basseterre riots a detachment of marines traversed the two-mile channel to neighboring Nevis. Nocturnal cane fires also had erupted there, apparently lit in sympathy with the visible blazes on St. Kitts, which had acted as signals or beacons suggesting similar incendiarism on the smaller island.[37] On Nevis fires broke out at Pinney's and Paradise Estates as well as near the hospital in Charlestown.[38] On Antigua particular notice was taken of the St. Kitts riots, where "hundreds of acres of cane [were] destroyed by fire, and we understand threats to burn down Basseterre have been attempted."[39] Fire and the threat of fire had, in other words, helped the events in St. Kitts take on regional, not simply local, importance in the northeastern Caribbean. For months thereafter, officials in St. Kitts, Antigua, and Montserrat took grave notice of any acts or even threats of incendiarism. One can only imagine their dismay, only a year later, at George Baden-Powell's cheery suggestion about "beacon bonfires" for the anniversary of Victoria's reign.

On St. Kitts itself the direct sociopolitical effects of the fiery riots were felt for at least a year. On the morning after the riots, a group of the island's leading planters and merchants met and called for the local formation of a "permanent protective force" as well as the continued presence of a warship in the Basseterre roadstead. The *Cordelia* soon was relieved by the HMS *Tartar*, which stayed until March, a necessity in official eyes because "the blacks are making threats to burn the town and murder the white people at the first opportunity."[40] A special circuit court in St. Kitts convened in mid-March to hear the cases of the accused rioters and convicted forty-three of seventy-four tried. By the end of 1896 the special protective force had become a reality and included a mounted company composed mainly of planters and merchants. Yet the formation of this armed body did little to prevent the rumors and threats of retaliation by black malcontents. A series of sugarcane fires on the island in November led the local administrator again to call for help. In late December the HMS *Intrepid* arrived to maintain order. The administrator noted in his correspondence to the ship's commander that, although the overall situation in St. Kitts had improved greatly since a number of rioters had been imprisoned, there still remained "a large number of idle vagabonds who have nothing to lose but possibly everything to gain by fire, riot, or disturbance."[41]

In reporting about the events in St. Kitts, the *New York Times* attributed the disturbances to the region's economic malaise and suggested that nearby British West Indian colonies might also erupt at any time for the same reason. The article claimed that the riots originated with a "workingman's strike in the capital" and then spread to the countryside—the exact opposite of the descriptions found in the numerous official reports. But in emphasizing the coalescence of town and country residents in creating and intensifying the disturbances, the New York paper seemed to underline the importance of large crowds themselves and the destructive directions they could take: "The rebellion appears to have been quite a spontaneous outbreak, without any organization and unpremeditated."[42] None of the reports of the St. Kitts riots enumerated or estimated the size of the "throng" or "large crowd" that marched through Basseterre on February 17, 1896, but they all found significance in the spontaneous violence that was an end result of the combination of rural workers and urban laborers uniting against common, even targeted, foes.

The St. Kitts riots thereby took on hemispherical importance, not simply significance confined to the British West Indies or the London Colonial Office. To the latter, however, the events in St. Kitts had a special meaning. For more than a decade, according to those in the Colonial Office who monitored events in the region, there had been reported rumblings from the working classes about riots and even "risings." An outbreak of fires of anger and passion was a sure sign of real trouble in any of these places. So when the riots of February 1896 occurred, they went far in inspiring the formation of the momentous British Royal Commission of 1897.[43] The commissioners met in London early in the year and then traveled throughout the Caribbean; based on their tours and local investigations, they eventually made pivotal decisions with far-reaching consequences. On one of the first days of 1897 the commissioners gathered information in London as a prelude to traveling to the Caribbean colonies. One of their interviewees, Alexander Crum Ewing, the owner of sugar-producing properties in British Guiana, Jamaica, and the Leewards, provided a lengthy technical narrative concerning soil fertility and sugar manufacturing. During his businesslike deposition, Ewing made the following direct observation: "The cause of the late riots and the manner in which they were dealt with are matters deserving the attention of the Commissioners; incendiary fires broke out again a few months ago."[44]

"Wrapped in Flames and Entirely Gutted": The Trinidad Water Riot of March 1903

The momentous civil disturbances in Port of Spain, Trinidad, on Monday, March 23, 1903, were described in retrospect by the London Colonial Office as "the most serious . . . in the West Indies since the Jamaica rebellion of 1865."[45] A violent extension of the decades-long public altercations over the city's water supply, the Port of Spain "water riot" possibly was misnamed, because the outbreak culminated in a major fire. In the early afternoon of March 23 a rebellious, stone-throwing crowd burned the Trinidad Red House, the long-standing architectural symbol of British colonial rule not only for Trinidad but for the southern Caribbean in general. Official retaliation led to the killing and wounding of scores of protesters; the official response furthermore involved the mustering of extra police, special troops (including those stationed in Barbados), and civilian reinforcements. Commentary and commissioners' findings in the aftermath highlighted resentments based on race and, especially, local frustrations centered on the arrogant and poorly administered ineptitude of Crown Colony rule in Trinidad. Repercussions from the riot reverberated throughout the region, especially in the smaller southern islands tied closely to Trinidad by intertwined economic, cultural, and demographic strands. The written commentary thereafter threw light on a number of the interrelated material and social problems that beset Trinidad at the turn of the century.[46]

In the week following the mass public meeting on March 14, 1903, in which various speakers described the government's water policy as tyrannical and ineffectual, a remarkable antigovernment solidarity united disparate urban factions of Port of Spain. Governor Alfred Moloney and his advisors nonetheless pressed forward with plans for a public reading and consideration of the new water bill, with its provisos calling for, among other things, metered water. Amidst a passion-suffused backdrop of escalating newspaper rhetoric and noisy public gatherings, the bill's reading was scheduled for the early afternoon of March 23 at the Red House, the colony's seat of government located at the city center between the police barracks to the west and Brunswick Square to the east. Anticipating a throng of onlookers for the hearings and citing a limited seating capacity, government officials published notices that those wishing to attend the debate would need to obtain printed tickets—consistent with hearings of the imperial parliament in London—that would be available at the door. By late morning on the twenty-third a crowd estimated in the hundreds and led by those who had spoken out in recent rallies attempted to enter the government chambers, claiming that members of the general public should be allowed to observe the deliberations with or without tickets.

At that point the commander of the local police force, Lieutenant Colonel H. E. Brake, attempted personally both to enforce the ticket rule and to restrain the increasingly boisterous crowd by deploying the police he had stationed around the building earlier in the day.[47] As the debates began, what, in retrospect, impressed Brake most was the incredible noise produced by the restless crowd outside the building. They were "singing, cheering and beating oil tins and drums, and the speakers in Council were frequently inaudible."[48] But what most recall about the unfolding of events was that the crowd suddenly turned violent in the early afternoon, hurling stones through the building's windows. The crowd then forced its way into the legislature's chambers, causing most of the officials to seek refuge. The angry, emboldened throng also tried to dismantle Governor Moloney's carriage, and some even considered killing the horses; several men eventually dragged the damaged vehicle to the city's waterfront and threw it into the harbor.

The riotous behavior was "blown into flame" about 2:30 P.M. The igniting of the Red House fire, however, was apparently less an instance of a riot that inevitably escalated into a fire than it was a clear case of arson, because afterward fire inspectors determined that, once inside the Red House, anonymous incendiaries lit fires, more or less simultaneously, in the registrar-general's office, the hall of justice, and the survey office.[49] Once burning, the fire spread rapidly, fanned by a wind from the southeast blowing through the open windows. The blaze raced through the flammable wooden interior, fueled by the furniture inside the Red House, and soon "the entire building, from end to end, was wrapped in flames and entirely gutted." The rioters appeared delighted by the conflagration, especially since not all of the officials had departed, and they "openly expressed their intention of roasting alive all within the Red House."[50] Serious injuries from the fire itself were, however, not reported afterward, although several officials were cut and bruised by stones. From all reports, the flames rushed through the building's interior, leading to the suspicion that the incendiaries may have used kerosene, but "of this there is no real evidence." Kerosene or not, there is little doubt that the presence of fire raised the riot's intensity to near-uncontrollable heights and helped to precipitate the events that followed: "With the spread of the fire, which in a few minutes reached the Council Chamber, the situation became highly critical, and immediate action had to be taken."[51]

Fortunately for the Trinidadian officials, a British naval detachment was in port. Captain C. Hope Robertson, from the bridge of the HMS *Pallas*, reported seeing smoke from the Red House fire about 2:00 P.M., the time when he received emergency messages—although the telephone lines from the Red House had been cut—to come ashore to assist, as "the lives of His Excellency the Governor, and other officials were in danger." Robertson then sent 156 bluejackets from the *Pallas* and the HMS *Rocket* ashore, along with three machine guns. The British naval forces, whose organization and discipline far exceeded that of the police, then dispersed some of the crowd, helped patrol the streets, and protected the other public buildings, including the Colonial Bank. By 6:00 P.M. all was reasonably quiet. Special cautionary measures accompanied the funerals of several of the slain rioters the following day, but little further trouble ensued. The British naval force occupied key points around the city and remained ashore until the morning of March 26. Then they were relieved by 200 British soldiers of the 3rd Lancashire Fusiliers, whom Captain Robertson had summoned from Barbados.[52]

Red House remains after the 1903 water riot in Port of Spain, Trinidad (from Gerard A. Besson, *The Angostura Historical Digest of Trinidad and Tobago* [Trinidad: Paria, 2001])

When the Red House fire was spreading rapidly, Colonel Brake sought permission from the colonial secretary to read the Riot Act. After receiving the authorization, he read the act from the veranda of the Red House amidst the fire's smoke and a shower of stones, bricks, and bottles. Brake then ordered riflemen to fire on a particularly hostile section of the crowd, which then broke and ran. Several policemen, apparently acting without orders, pursued fleeing members of the crowd south from the Red House, firing and using their bayonets. Order finally was restored. The Port of Spain water riot had taken a heavy toll in property. Beyond the destruction of the interior of the Red House, the top floor of the police barracks was set ablaze late on the same afternoon and then burned itself out, as there was no relief supplied from the "miserably small jets of water available."[53] A far heavier toll in human life saw 15 (9 men and 6 women) presumed rioters killed, some of whom died in the hospital among the 49 (36 men and 13 women) receiving treatment there for bullet and bayonet wounds. Doubtless more suffered minor injuries but did not come forward for medical attention for fear of being prosecuted as rioters.[54]

The riot, which focused on Port of Spain water issues and obviously took place in the city itself, nonetheless affected much of the rest of the island. Before and during the disturbances, Colonel Brake summoned police reinforcements from beyond the city as well, individual members of volunteer paramilitary units who were charged with both defending Trinidad from external aggression and maintaining the peace in times of internal disorder. Members of the "Light Horse, which consists of Europeans and Creoles of European parentage" helped to protect the Red House and the officials inside it during the water riot, as did individuals from the Volunteer Light Infantry, men with similar bloodlines and economic interests.[55] Earlier in the day, after some of the local police and light infantry volunteers left San Fernando by rail to augment the capital city's police force, some of San Fernando's "rowdy element" took advantage of the relative lack of security left behind. Inspired by the events in Port of Spain, the San Fernando roughs roamed the streets, threw stones, and threatened to set fires. Most ominously, some of them loudly suggested, "Let us kill the white people," thereby seizing "an opportunity to display the bitter hatred they entertain towards the white population."[56]

News of the water riot spread quickly, not only locally by word of mouth but also beyond Trinidad via telegraph. The conservative newspapers of the British Caribbean region, such as the planter papers of Barbados, reported the event as a destructive interference with the colony's orderly economic activity and praised the "invaluable service" rendered by the bluejackets who had debarked to preserve the peace from the actions of the "riotous mob."[57] Residents of the smaller islands read a far different interpretation of the incident, in which the Trinidad officials were bitterly condemned not only as the perpetrators of the riot but also for the way they put it down. Those in St. Lucia, for example, read of "a revolt against arbitrary use of power . . . by a bitter sense of injustice, which has bespattered the walls of Port-of-Spain with blood and strewn her streets with corpses."[58] After the outbreak of riots was initially reported in the New York and London newspapers, follow-up stories in both places underlined the thorough unpopularity and apparent incompetence of Trinidad's political administration by pointing out that the local chamber of commerce had adopted a resolution asking the Colonial Office to remove the governor and the island's top officials "in whom the public has entirely lost confidence."[59]

Beyond the informal hearing aired in the press, the Port of Spain water riot led to two different government investigations. A commission of enquiry called by the Colonial Office gave invaluable perspective on the origins and sequence of events. The investigation involved three commissioners who sailed from London to Trinidad in the latter half of April 1903. They departed Port of Spain on May 22 after interviewing 146 witnesses in Trinidad and visiting the scenes of the disturbances. The commissioners also assessed the background and current state of Port of Spain's water supply, an investigation that included tours of the city's reservoirs and pumping stations.[60] Then in August 1903, three months after the departure of the British commissioners, Henry Bovell, the chief justice of British Guiana, came to Trinidad and interviewed 295 witnesses in the course of his investigation of the "excessive and uncalled-for shooting and unjustifiable use of the bayonet by the Police on the 23rd March last."[61]

Commentary sent directly to London from the Chamber of Commerce and the Ratepayers' Association emphasized the government's inability to maintain a water supply sufficient to combat fire in Port of Spain. They placed particular emphasis on the water inadequacies during the March 1895 fire, which should have alerted authorities to take appropriate action. Then London's attention was drawn to the March 1902 fire at the Trinidad Shipping and Trading Company, only a year before the water riot, during which a water shortfall similar to the one in 1895 contributed to massive property destruction. As far as the ratepayers were concerned, the evidence could not be clearer. They called for the immediate removal of the local officials who had deprived the community of proper fire protection. Modern problems called for enlightened leadership, not the ineffectual British colonial bureaucracy that had been reduced—to hear the Port of Spain ratepayers describe it—for all peoples everywhere in the empire except "for us official-ridden Trinidadians."[62]

In describing the antigovernment rioters—and presumably those who lit the fires—at the Red House, the published, post-disturbance commentary was at once predictable and revealing. At the predictable level, official generalizations pointed out the inferiority and excitability of those who created the disturbances and hurled rocks. They were the "rowdy elements" of Port of Spain, "the lowest class of coloured people—thriftless and lazy, who live by doing odd jobs about the harbour and the like."[63] Yet a closer look shows that these same people were not mindless malcontents bent on trouble and riot for its own sake. Rather, they had followed closely the political events surrounding the city's water supply by, among other means, reading about them in the local press: "The poor common people . . . most densely ignorant, have no general ill-feeling and when unexcited are docile and kindly. But they are terribly subject to be excited into acts of violence, and are ready to accept as true . . . any statement made in language intelligible to them, spoken, or, *still more, printed*."[64]

In the aftermath of the riot, various commissioners attempted to recreate the events with an eye toward fixing responsibility for the loss of life and the breakdown of civil authority. In assessing what took place on March 23 and immediately thereafter, they found a troubling overlap in lines of authority and a wide array of individual performances under fire, ranging from heroism to cowardice. The fire brigade, cursed with an ironically thin water supply, did little to distinguish itself otherwise. In testimony afterward, Colonel Brake professed surprise that the crowd burned the Red House, yet he must have suspected that possibility as early as two days before the riot because he notified Captain Darwent then to be prepared and told Darwent on Sunday to have plenty of canvas hose available. But when the burning began on the afternoon of the twenty-third, the angry crowd directed its rage toward the fire brigade barracks with "a series of volleys and showers of stones, and defied anyone to turn out."[65] At that point Darwent, who was afterward dismissed from government service as a result of his performance, apparently hid inside the fire brigade headquarters. When authorities sought help from the fire brigade at a crucial point, they were informed that Darwent was "ill in bed."[66]

Those who fired rifles from the Red House into the midst of the crowd were white Trinidadian volunteers. As the March 23 showdown between protesters and government officials approached, Colonel Brake assembled a motley force of roughly 200 police and volunteers. Some armed with rifles were stationed inside the Red House before dawn on the day of the riot. Just before the crowd approached the government buildings in the late morning, 64 police were sent to the Red House armed with "long staves," and then another 25 came from the police barracks a little before 2:00 P.M.[67] After the demonstration erupted and the Red House was set afire, Colonel Brake, after reading the Riot Act, led a dozen or so riflemen to the eastern side of the Red House and instructed them to fire into the crowd. "Included in the armed party were three civilians, whose services had been offered to, and accepted by, Colonel Brake earlier in the day. These were Messrs. Hubert Patterson and Colin and Reginald Harrigan. All three were clerks in the Government service, and the two former are also troopers in Captain Bowen's troop of the Light Horse." Patterson and the two Harrigans were among the riflemen firing from the Red House veranda, and then the others reached the ground, "spread out, and fired into the crowd at close quarters."[68]

Brake reportedly urged the troops to clear the protesters from the grounds of the Red House, but a few of the overzealous black police exceeded their authority, chasing rioters south along St. Vincent Street and firing into the crowd. Two of the females in the crowd, Millicent Hadaway, a twenty-three-year-old cook, and twelve-year-old Eliza Bunting, were killed with bayonets. But a trial early in 1904 of a police sergeant accused of using the bayonet came to nothing, mainly because of a dearth of witnesses willing to come forward; "one witness . . . described having seen a policeman bayonet a young girl . . . but his evidence was confused and contradictory."[69] The police brutality was thereby compounded—and perhaps underlaid—by a lack of trust between the people on the street and island authorities. Similar mistrust led to a complete absence of prosecutions for those who had ignited the government building. After the water riot, Colonel Brake commented as to why no one, given the hostile atmosphere in the city, would provide evidence of arson. Although, according to Brake, the setting of fires had to be known to "dozens of persons," it was, however, impossible for anyone to come forward. "It is perfectly certain that any common person giving evidence in this matter would be in considerable danger and this fear has undoubtedly much to do with the conspiracy of silence which has been maintained."[70]

A remarkable dimension of the 1903 Port of Spain water riot is that, in terms of the identities of those wounded and hospitalized, the riot was a regional, not necessarily a Trinidadian, affair. Of the 49 persons hospitalized from injury—and presumably representing a cross section of the 5,000 or so rioters—35 were from outside the island: 15 were from Barbados; 8, from St. Vincent; 1, Montserrat; 1, Haiti; 1, Ireland; 1, Venezuela; and 8, unknown. Some of the latter had names that sounded suspiciously Barbadian, such as Greaves and Yearwood.[71] These data confirm that members of the black underclass of Port of Spain—and perhaps those who were the most socially active—represented many who had come to the larger island in the previous decades. Lacking the social support that an extensive family network on the island might provide and without the benefit of a traditional means of coping with material shortages or government indifference, these new immigrants likely had few alternatives to solving problems other than confronting local authorities to improve their well-being. In any case, these astounding hospital data underline the importance of the observation made by Trinidadian historian Bridget Brereton about the composition of the Port of Spain underclass in the latter decades of the nineteenth century: "In the last thirty years of the century, the black and coloured masses were being increased year by year by an influx of immigrants from the eastern Caribbean and especially from Barbados."[72] One can only begin to imagine the hundreds of village-level conversations throughout the eastern Caribbean at the time about the Red House fire and the water riot.

In July 1903, three months after the water riot, a letter from Joseph Chamberlain, the secretary of state for the colonies, to Trinidad's Governor Moloney included observations about the event and recommendations for the future. Aghast at the "conflagration," both figuratively and literally, that the water riot represented, Chamberlain insisted that no further delay be taken in attempting to identify who was responsible. In reviewing the performances of various individuals and units, he singled out the fire brigade as "deplorable in the extreme" and recommended it be placed directly under police supervision. Trinidad, thought Chamberlain, had a tradition and reputation "for the leading and most respected members of the community to be usually in opposition to the Government." Yet this apparent "cleavage between the rulers and the ruled" struck Chamberlain as contradictory to the colony's economic success; if this rivalry was truly as pervasive and widespread as advertised, the island never could have become so "conspicuously prosperous."[73]

"The Country Districts . . . All Ablaze": St. Lucia in April 1907

In late April 1907 an entirely surprising yet exceptionally damaging mass protest shook St. Lucia and, indirectly, nearby island colonies that provided extra police and volunteers to help quell the disturbances. The St. Lucia incident began with seemingly innocuous petitions to the new governor by the male and female coal carriers in Castries for higher wages. After he responded to what he considered a minor request, rural sugarcane workers from the valley area south and east of Castries marched on the town, demanding better pay. Islandwide protests then exploded and lasted for several days, featuring, among other things, various groups invading Castries and looting the marketplace and selected stores and shops. These were altogether shocking disturbances "due to one of those sudden and unpremeditated waves of passion which have sometimes swept over communities."[74] The culmination of the riots saw sugarcane burned from the Cul-de-Sac factory south of Castries all the way across the island to Dennery on St. Lucia's windward coast. Police shot and killed four and wounded nineteen at the peak of the riots on Thursday, April 25. Several officials and business leaders were injured during the disturbances. The ranks of local police and volunteers were augmented with personnel and weapons from Barbados and St. Vincent as well as a welcome (as far as British colonial officials were concerned) and unexpected contingent of Dutch marines.

Unlike in the case of the Trinidad water riot four years earlier, which capped a seemingly interminable dispute over public resources, St. Lucia officials had no idea disturbances were coming. Almost as soon as the new governor of the Windward Islands, Sir Ralph Williams, was sworn into office in April 1907, he was approached by small groups of "coal heavers" off the St. Lucia wharf when he visited from his headquarters in Grenada. The coal workers apparently had taken seriously the governor's overtures in his welcoming address that anyone, regardless of his or her rank or station, could approach him to air grievances. In the afternoon of Friday, April 19, some workers left their jobs off-loading coal from two European steamers at the Castries harbor to complain to Williams of low pay, the colony's onerous road tax, and the hasty burial of paupers in makeshift caskets. The small group requesting better overall conditions grew into a large crowd by mid-afternoon. Backpedaling from his earlier benign openness, Williams explained through an interpreter to the mainly Creole-speaking crowd that he had no control over wages, that laws of economics—certainly not politics—dictated earnings, and that the men who owned the island's coal facilities (who, everyone knew, lived in opulent homes overlooking Castries) certainly would pay more if times were better. Few workers returned to their jobs on Friday afternoon, and the streets of Castries took on an impromptu carnival atmosphere, with a good deal of dancing, singing, and waving of branches. Similar festivities and public displays continued into the following day, and Sunday sermons in Castries were devoted in part to convincing the faithful to return to work the next day.[75]

The island's real upheaval began on the following Tuesday. Governor Williams left St. Lucia for Grenada via Barbados aboard the steamer *Esk* on Monday morning. The mood among the crowd at the pier was "not by any means as cordial as that which greeted his arrival." On Tuesday the coal carriers refused to refuel the British mail steamer *Solent* when it arrived that morning from Panama. Four women who attempted to load coal were attacked by others and needed a police escort into town. A growing crowd then chased the women into a store, looted it, and began to single out white merchants and businessmen, beating them and then attacking and destroying their homes. As the violence in the streets escalated into a general disturbance, some forty "gentlemen"—probably all, or nearly all, of whom were white—were sworn in as special police constables to complement the regular police force. Several of these special constables, including W. M. Howell, the manager of the Colonial Bank, whose leg was broken, were then badly injured by rocks and sticks wielded by dockworkers who had been joined by others. Ten policemen armed with rifles were cornered by a large group of protesters and opened fire, injuring four. By evening several of the streets were strewn with goods that had been taken from the local stores and shops, although the disturbances did not continue into the night.

Notified by telegram while aboard ship, Governor Williams returned to St. Lucia about 3:00 A.M. Wednesday. Once ashore, he headed for the central police station, where the special constables were waiting, and he discussed the events with a "number of responsible persons" while telegraphing the Colonial Office for help and Barbados for extra volunteers. Shortly after dawn, Williams and his secretary conducted a personal walking tour of parts of Castries, exhorting the people to cease making trouble and to get back to work. Upon his return to the police station, Williams was informed that rioters were ransacking the city market. Off he dashed to the marketplace, pushed his way through a noisy crowd of protesters brandishing cutlasses, climbed atop some boxes, and successfully ordered the crowd, through an interpreter, to calm down. In his memoirs Williams considered this event pivotal during the St. Lucia episode because at that point someone easily could have struck him down. Had that happened, in Williams's estimation, "I am certain that within half an hour there would have been a massacre of every white soul in Castries and later on in the whole island."[76] In any case, while the governor was attempting to restore order at the Castries city market, the coal employers agreed on a higher wage scale. News of the tumultuous events in town then raced through the countryside, and soon "the country districts . . . were all ablaze."[77] By midafternoon a crowd of angry laborers from the Cul-de-Sac sugar factory two miles south of Castries marched on the town for better wages, but their collective wrath was defused by the reluctance of the coal workers to join them, as well as by the governor's invitation to several individuals in the group to walk freely through the town.

Thursday saw the height of the St. Lucia disturbances. Adding to, and probably precipitated by, the troubles in town, gangs of malcontents burned patches of sugarcane throughout the country districts, thereby raising tensions to new heights. In the early morning, news reached Castries of riots and burning at Roseau, four miles farther inland from the Cul-de-Sac factory. A police detachment headed for Cul-de-Sac, where a crowd of cutlass-wielding laborers estimated at 2,000–3,000 from the two sugar complexes had congregated; they had destroyed gardens and slaughtered draft animals, and the cane was ablaze. The police and rioters exchanged threats and counterthreats, the crowd advanced, and the police fired, killing four and wounding a score of others. Governor Williams arrived shortly afterward, and later he described the corpses he encountered from the shootings. He then traveled on horseback to Roseau—accompanied, remarkably, by some of the rioters—where he found the sugarcanes ablaze and some of the factory buildings soaked with kerosene and ready for burning. As he had done at the Castries marketplace, Williams brazenly cajoled the crowd into more peaceful behavior and even persuaded some of the rioters to help put out the fires at Roseau. Later on Thursday, yet another group of rural workers, from Dennery on the windward side of the island, marched on Castries with "rude flags and rough music." Williams later described the white inhabitants of Castries at this point as thoroughly terrified. The rioters and marchers, in the meantime, referring among themselves to the nearly empty harbor, saw little to prevent them from murder and plunder, boasting openly in the streets of Castries, "There are no English Men of War now."[78]

External help for the beleaguered whites in St. Lucia came none too soon. A Dutch warship, the *Gelderland*, arrived on that Thursday afternoon, while the Dennery laborers were parading the streets of Castries. The Dutch vessel and its contingent of marines were seeking coal supplies they had been unable to find at Martinique. On Friday morning the *Solent* returned from Barbados, this time carrying the Barbados chief of police, Colonel Kaye, and ninety police and volunteers along with a machine gun. Soon thereafter a sailing sloop with more police aboard arrived from St. Vincent. On Friday morning Governor Williams rode his horse across the island to the Dennery sugar factory only to find that the town, about three miles distant, was being looted. He and a small detachment of police, with the moral support of the local Catholic priest, attempted to quiet the disturbances there. Yet more police from Barbados arrived in the afternoon at Dennery aboard the coasting steamer *Tees* and began to arrest some of the men accused of looting and burning.

The reinforcement of the local police force, whose ranks had been well augmented with external police and volunteers, allowed the governor to ride back to Castries at a leisurely pace; the worst of the disturbances finally were over. The *Gelderland* nonetheless remained in the harbor until Monday, April 29, 1907.[79] During the weeklong St. Lucia disturbances, newspaper readers in London had been well informed. Articles appearing there stressed the wreckage of sugarcane factories and the destruction of crops by burning, as well as the comings and goings of various vessels carrying police reinforcements.[80] A week after the Dutch warship departed Castries, the attendees of a mass meeting of legislators, merchants, and other "prominent inhabitants" of the town considered and approved the formation of a local volunteer corps. They also discussed the idea of dividing St. Lucia into various districts for better control and policing. As in other islands—including St. Kitts after the 1896 riots there—those attending emphasized the importance of using horsemen to form a mounted cavalry of volunteers.[81]

Even if Governor Williams's descriptions of the events in St. Lucia were exaggerated, he nevertheless displayed extraordinary boldness and bravery and took great risks at several points by taking charge personally. His stated belief for doing so was that the lives of all white St. Lucians were at stake, a point he reiterated several times in his published memoirs. These dangers were, in his estimation, by no means limited to St. Lucia, because he considered the April 1907 riots symptomatic of what likely might happen elsewhere in the "negro islands" of the Lesser Antilles, where, unless better protection from the British navy were made available, "there will sooner or later be a catastrophe of magnitude."[82] In the particular case of St. Lucia, Williams considered the arrival of the *Gelderland* pivotal in turning the tide. On the Friday morning before he rode across the island to Dennery, Williams visited the ship's captain, who assured him that all would be well in Castries. Some of the Dutch officers even expressed enthusiasm over the chance to "have a go" at the rioters; Williams in turn felt a "kinship" with the "great fair-haired and burly bluejackets" aboard the Dutch vessel, whose presence "blotted out all soreness over the events of the South African war."[83] As in the relief and immediate containment of other natural and man-made catastrophes of the eastern Caribbean at the time—including the hazards and dangers of the great city fires—nationality therefore seemed to play a far less important role than did the maintenance of overall metropolitan control. Put in a more straightforward, and almost certainly a more truthful, way, nationality meant little when it was a case of white versus black, the overriding sociodemographic fact of the British Caribbean region.

The surprise expressed by the St. Lucia authorities over the 1907 troubles doubtless stemmed from the mutual misunderstanding among the members of the different cultural divisions on the island. Although St. Lucia was nominally a British colony with a cadre of white, English-speaking officials, most of its inhabitants were rural blacks who spoke a French-based Creole. The latter made up most of the laborers in the rural sugar factories, and their sympathies probably were in line with those of a small group of white, French-speaking planters and Catholic clergy whose loyalties to the Crown the British counted as tenuous at best. In Castries a visible, vocal, and transient element of English-speaking Barbadian and Vincentian blacks was preferred by the local authorities for coal yard and other dock labor activities. The mainly Barbadian police, moreover, generally were despised by the Creole-speaking blacks of St. Lucia, as they were in some of the other islands.[84] Although Governor Williams and other officials—all of whom delivered their public pronouncements through interpreters—did not understand their St. Lucian citizenry as well as administrators on other islands where everyone more or less shared a common language, they certainly knew enough to sense trouble if ever city and country people coalesced against a common foe. When Williams, for example, rode to the Roseau factory, and away from town, on the Thursday of the disturbances to hear workers' grievances, his real purpose was to keep country and city people from uniting. His overall strategy in trying to lead the Roseau workers back to their part of the island was, as he later recalled, "to break the combination."[85]

Beyond the fundamental socioeconomic differences between blacks and whites common to and underpinning the civil disturbances at the turn of the century in St. Kitts, Trinidad, and St. Lucia, perhaps the most general common denominator was that all three were animated by fire. The three were obviously grouped together by geographical location in the minds of the men at the London Colonial Office and thereby were considered part of an overall British West Indian problem. But a strike and subsequent rampage by a sugarcane proletariat in St. Kitts, the explosion of a long-simmering dispute over public resources in Trinidad involving sophisticated antigovernment spokespersons, and the wage revolt by an enigmatic and linguistically alien working class in St. Lucia bore little resemblance to one another. What they did have in common was a riotous tone and a damaging and frightening incendiarism that carried disgruntled reaction at the local level onto a completely different and near-uncontrollable plane. Fire scared powerholders, and everyone knew it. Furthermore, it was the ideal vehicle for revenge. Early in 1905, two years after the water riot and two years before the disturbances in St. Lucia, the new governor of Trinidad wrote a lengthy memo to London. In it he pointed out that he had learned what certain blacks were preparing to do during the forthcoming carnival season now that the British soldiers no longer were in the colony. Complaining of the onerous and ongoing taxation policies of the government, and referring self-consciously to the fire they planned to use against the "white people," black spokesmen had been very specific. They were preparing to "make all the Government people 'smell hell' " during the carnival parades.[86]

8

Epilogue

GOVERNOR RALPH WILLIAMS'S EXPRESSIONS of fearful concern for the overall safety of whites residing in St. Lucia during the troubles in April 1907 spoke directly to London's recent decisions to withdraw nearly all of the British military forces from the Caribbean region. These decisions created regional, not simply local, problems, and Williams was echoing similar apprehensions from all of the islands, each of which had been affected—as far as local white residents were concerned—in the same way. In the midst of the 1907 St. Lucia disturbances, the local newspaper had emphasized that the St. Lucia troubles really reflected, at the microscale and on the ground, "the results of the [British navy's] bluewater policy, and the practical abandonment by Great Britain of the West Indies to their casual fate."[1]

This "practical abandonment" referred to Britain's strategic removal of its armed forces from the British West Indies two years earlier. The withdrawal, moreover, was an example of the kind of external decision making, far beyond local control, that always has affected the Caribbean region and each of its particular islands. From about 1880 to 1905, geopolitical conditions had changed substantially in the British Caribbean and elsewhere. Late in the nineteenth century Great Britain's North American and West Indian Squadron had represented the empire's active naval presence in the Western Hemisphere. The squadron's annual sailing circuit—linking Halifax and Bermuda with the British Caribbean ports all the way south to British Guiana—was based on the acknowledgment of the United States as its chief military rival in the region. Local residents of British Caribbean islands almost certainly interpreted the late nineteenth-century British naval presence there in a more immediate way. If local troubles ever were to occur—as they occasionally did—a telegraphed message would bring a British ship, its deck brimming with armed bluejackets, to the local harbor in a matter of days or even sooner. This probability, as protective and reassuring to white officials and planters as it was forbidding to those intent on creating disturbances, represented an important element in the overall colonial control, and almost certainly the daily social atmosphere, of individual British West Indian colonies.

Yet as Britain's geopolitical interests shifted away from the Western Hemisphere, partly in response to the growing U.S. supremacy in the Americas, major changes in the deployment of British armed forces in the Caribbean region became inevitable. Early in 1905, as a result of the secret discussions in London by the Committee for Imperial Defence that had been ongoing for two years, the British government abolished the North American and West Indian Squadron and announced the imminent withdrawal of almost all British soldiers from its Caribbean possessions. Officials from the Colonial Office, who were of course directly responsible for the day-to-day operations of the empire's individual colonies, had, for all practical purposes, been excluded from these discussions from the start. When the news of troop withdrawals triggered the predictable protests from the planters and merchants in the West Indies, these misgivings were transmitted by the Colonial Office to other government authorities in London, but they seem to have had little overall effect. Cries of alarm from small, anachronistic island colonies apparently meant little in the face of shifting imperial strategies in the early twentieth century. Sir John Fisher, the first sea lord, who recently had commanded the North American and West Indian Squadron, responded airily to the pleas for an ongoing presence of white soldiers in the British Caribbean that neither the War Office nor the Admiralty could spare valuable personnel and equipment simply to indulge in "police work."[2]

The exodus of European soldiers from the British West Indian colonies seemed particularly ominous, to say the very least, to local whites. Early in 1905 the British colonial secretary, Alfred Lyttleton, sent a circular letter to all of the Caribbean governors informing them that white infantry would leave the region "as soon as possible," including the detachments at Barbados, Jamaica, and the coaling station at St. Lucia. A small company of the (mainly black Jamaican) West India Regiment would replace the British troops in Jamaica. Lyttleton acknowledged that the white officeholders and

Departure of British troops from Barbados, November 1905 (from Ann Watson Yates, *Bygone Barbados* [Barbados: Blackbird Studios, 1998])

businessmen throughout the region would be distressed, but there was little the Colonial Office could do except to advise the locals to provide for their own security, as some of the island colonies already had done during the disturbances of the 1890s. Lyttleton concluded the message by reaffirming that "each colony must rely in future on its own resources for maintenance of law and order."[3]

The fearful reactions from the white residents of the region took a number of forms. Beyond what has to have been the main discussion topic among friends and family members for months, written responses were impassioned and widespread. In London the West India Committee pointed out the "grave risks," in light of recent disturbances from the Leewards south to British Guiana, of the presence of volatile "labouring classes . . . whose excitable nature is well known, and whose numbers are so enormously in excess of the white inhabitants." If recent history was any guide, the committee's statement continued, the explosive force of riots and disturbances that this demographic imbalance would create simply would be "beyond the powers of the ordinary police force to cope with."[4] The *Barbados Agricultural Reporter*, typical of most newspaper reaction throughout the region, saw in the recent past the basis for a troubling—indeed fiery—future: "We are told that we must rely on Volunteers and police. Did either . . . avail when disturbances arose in Trinidad, at as late a date as March, 1903, when the Red House was burnt down?"[5]

Yet the rebellious, vengeful bloodbath, so widely predicted by Governor Williams and others as the inevitable aftermath of the removal of a permanent military force, never took place in the British Caribbean. As an obvious corollary—although musing about the factors that help explain why an event never occurred can approach pointlessness—one must acknowledge that the white officials and residents in the region at the turn of the century, given their widely aired fears, simply misjudged the mood and character and culture of the people they dominated.

This misjudgment likely increased as a function of distance from the West Indian islands themselves. Whereas there were those in London who understood some of the colonies abroad by virtue of their personal experiences, at the highest governmental levels in London, ignorance of local conditions and moods in the empire obviously was profound. Colonial Secretary Joseph Chamberlain, to cite a most relevant example, found the 1903 Trinidad water riot disquieting and bothersome. The hundreds of documents sent to London in the months prior to March 1903 demonstrated clearly that the conditions leading to the water riot revealed a dispute between a vociferous, increasingly sophisticated, and genuinely broad-based local group intent on improving their own living conditions versus a bumbling colonial administration. But to confidants Chamberlain characterized the flaming outburst in a far different way. It was, he suggested, symptomatic of a reaction by near-savages who did not appreciate the benefits bestowed upon them by the empire: "I have an idea that we press sanitation & civilisation too strongly on some of these backward communities. If they like bad water or insufficient water it might be well to let them find out the results for themselves."[6]

Although West Indians were unaware of the colonial secretary's remark at the time, buried as it was in a confidential ledger, certainly they sensed—and resented—the condescending arrogance with which the British authorities misread the aspirations of them as British subjects. Emmanuel Lazare, it may be recalled, had had to remind the queen herself that the residents of Trinidad, while being Trinidadian, were culturally English to the core. If Port of Spain were indeed a backward community, as the colonial secretary had later suggested, London had only itself to blame, because as every Trinidadian knew, the island's infrastructure, population, languages, and all else were, in many ways, colonial creations.

These local feelings of resentment played a part in and could perhaps be said to help underpin the regional disturbances at the turn of the century—feelings that had been thrown into high relief and exposed by, among other things, the many fires of protest. Grassroots commentary about, again, the Port of Spain water riot provided vivid examples of these feelings, resentments felt not only in Trinidad but throughout the smaller islands of the region as well. In commenting on the sending of locally housed British soldiers to Trinidad to maintain order there after the water riot, one Barbados newspaper pointed out that the problems lay not in the rebellious nature of the people of Trinidad but in the appointed colonial leaders there who insisted on treating Trinidadians as if they were "natives."[7] The Grenada newspaper raised the same issue, but in more graphic terms. It pointed out that Port of Spain police superintendent Colonel Brake (of water riot notoriety) had earlier served the empire as a military officer in Africa. In describing the use of firearms and bayonets by Brake's police in Port of Spain, the Grenada newspaper exclaimed, "It appeared to us to be a punitive measure, an act of revenge on the people for having set fire to the Red House. . . . Can an officer be allowed with impunity to treat His Majesty's subjects in any part of the Empire as he has been accustomed to treat the savages in Central Africa?"[8]

The majority of West Indians in the early 1900s most likely bristled at the thought that they were being treated like "savages." Nearly everyone in the region recognized the value of education, if for no other reason than as a practical way out from plantation labor for themselves or their children. So literacy was growing apace. As only one of many results of these trends, the increasing number of black and brown men in professional positions provided educated leadership for local causes. As Trinidad historian Keith Laurence has emphasized in quoting a memoir composed at the time: "Education, the arts, and whatever contributes to western culture find an easy home in the West Indies, the progress of which has no need to lower its head when compared with other lands."[9] More recently, Stephen Kingsley Scott has pointed out the importance of the attributes of conventional English culture—education, professed religious faith, Christian marriage, the proper use of English, and so forth—for success and self-esteem among the upwardly mobile black and brown residents of urban Trinidad in the late nineteenth century. At the same time, these same native Trinidadians were highly critical of the political arm of the mother country for sending maladroit colonial bureaucrats to rule over them.[10] Although Scott's work is limited to Trinidad, there is every reason to assume that these same ideas extended to other parts of the British Caribbean.

The cross-cultural comparisons at the time, considering the inhabitants of the Caribbean alongside those of Africa or other so-called savage or backward areas of the empire, are less important here than acknowledging the overall misreading of West Indians by the London Colonial Office. The misreading, moreover, reflected a profound and growing divergence in intent, interest, purpose, and goals between the British Colonial Office and its subject West Indian colonies. Generally, the great majority of appointees sent by London to govern the region saw control and order as their principal goals. Those native to the region, on the other hand, were involved in incipient nation building. Accordingly, they were attempting to create better lives and opportunities for themselves and their families in the only countries they had ever known, and with a growing fund of knowledge at their disposal. Riots and disturbances, although they could get out of hand, could eventually lead to positive results by convincing London to grant greater power and latitude to locals. Destruction, ruin, and disorder for their own sake, in contrast, made little sense.

The historical events created and inspired by these opposing viewpoints and aspirations unfolded, as they always have, within the Caribbean's physical geographical context—on the islands and in the intervening channels and surrounding waters of the region itself. The departure of the British naval vessels and soldiers, given the Caribbean region's far-flung, hazard-prone insularity, provided perhaps the most obvious and immediate example of an intimacy between history and geography working together at the time. Local white critics of the military exodus argued and pleaded that the British withdrawal left not just single colonies but a series of island possessions undefended from the black laborers. The British conceded to these pleas, acknowledging the need for at least a minimal military presence in the region, and therefore provided a new cruiser to be stationed permanently at Bermuda, except during the Atlantic's hurricane season.[11] White residents of the British Caribbean, however, found little consolation in the decision; relying on a single vessel to protect a series of islands simply spread too thin the meager-to-nonexistent military resources over too wide an area. Their reservations were in part confirmed during the St. Lucia disturbances of April 1907. Although Governor Williams cabled London early during the impending riots to ask for naval help, the *Indefatigable*, anchored at British Honduras in the far western Caribbean, was the only vessel available in the region at the time. The *Indefatigable*—dubbed humorously by local wags the *Ungetatable* because of its usual lack of reliability—arrived at the Castries harbor several days late, well after the disturbances had died down.[12]

The *Indefatigable*'s late arrival in St. Lucia also reinforced the widespread regional concern—among members of all classes of people—that had been created by the far more momentous episode of a few months earlier. The earthquake in Jamaica in January 1907 had reminded everyone in the British Caribbean how the new order might be influenced by the region's physical characteristics. The earthquake had inspired newspaper headlines and sparked diplomatic controversy only partly because of the catastrophe itself. At least as important was the fact that Kingston had been left unprotected from local hazards owing to the loss of the British naval vessels a matter of months earlier. Nearly a full week following the Jamaica earthquake, a British correspondent, combining incredulity with outrage, asked rhetorically from his vantage point in Kingston what in the world had happened: "What are we to say of the British Admiralty and to those who have mismanaged matters that on the fifth day after such a disaster . . . not a single British pennant [is] to be seen [in] the Kingston harbour?"[13]

The reactions by West Indian writers elsewhere in the region—who had more at stake in the incident than British pride—pointed out that small, disaster-prone colonies owned by an imperial power unwilling or unable to provide immediate help when environmental catastrophe hit were better off seeking political affiliation elsewhere. Their comments apparently were not considered seditious, and they made sense. The *Antigua Standard* mused whether, under the circumstances created by the withdrawal of British forces from the region coupled with the earthquake, it might not be best for Antigua to seek some sort of political relationship with the United States.[14] The tempestuous Grenadian editor William Galwey Donovan commented bluntly about the British dismissal of the American navy whose vessels came to Kingston, days before the British arrived, in the wake of the disaster: "The entire incident suggests a valuable lesson. It points to the decay of British influence in the West Indies. . . . Governor Swettenham's action was reprehensible. He should have allowed the American marines to land, encamp on British territory and leave the work of rescue entirely to Admiral Davis."[15]

The remarks by Donovan and similar commentary from other black intellectual leaders of the region at the time would be badly misinterpreted if they were considered pro-American. Donovan, acquainted through correspondence with his readers with the labor conditions on the Panama Canal, for example, had to be acutely aware not only of American racism and oppression at the canal but also of similar conditions associated with Cuban sugar production and other American-financed projects in the region. American racism and gruffness accompanying a relief and rescue operation from a hurricane, earthquake, or major city fire was, however, preferable to little or no help from a British colonial office no longer willing or able to respond effectively. Donovan's and others' similar sentiments, therefore, were not pro-American so much as they were pro–West Indian. Within the context of these important historical and physical-geographical events, Britain's military withdrawal from the region therefore had had the unlikely effect of displeasing nearly everyone in the British West Indies, rulers and ruled alike.

The British military withdrawal in 1905 represented, among many other things, a lessening of the United Kingdom's physical responsibility for the

Caribbean region, yet London continued to control its Caribbean colonies through appointed and local white elites. In response, black and brown people of the region continued to resist this arrangement in a variety of ways, including work stoppages and strikes whose details were publicized through increasingly vociferous and articulate working-class newspapers. Returning West Indian veterans of World War I protested in various places in the region, and they were particularly effective in demonstrating a heightened capacity to organize. Two decades later a series of riots and disturbances in the 1930s were interrelated with the relaxation of the voting franchise from one island to the next. An increasing number of black labor leaders/politicians then led most of the small islands into the West Indies Federation during the 1950s. Political independence—or at least internal political autonomy—for all of the individual islands was granted starting in the early 1960s and extending into the 1980s.

The many forms and uses of fire—still as much an element of the Caribbean's environment as earthquakes, hurricanes, and insularity—accompanied the events surrounding the British military withdrawal and the other regional transformations of the early twentieth century. In many cases, fire's role in the islands in the early 1900s was much as it had been in the late nineteenth century and earlier, thereby reinforcing a continuity with the past. Illicit sugarcane fires were still reported, although not at the same level as during the turbulent protests of the 1890s. Shepherds continued to burn grazing areas, and the incidence of bush fires always increased in dry seasons. Small house fires owing to the improper uses of kerosene in crowded sections of cities and towns underlined one of the daily risks of urban impoverishment. City and town fires continued to be a dreaded menace, and descriptions of their occurrences were reminiscent of similar fires for several decades past. A fire in Grenville, Grenada, in March 1905, for example, destroyed several buildings of the town. The police were of little help, as they "smashed" three fire hydrants in attempting to affix hoses, but the general populace formed a bucket brigade and finally quenched the blaze.[16] Six months later, yet one more example of the tradition of international fire control in the Lesser Antilles saw visiting Venezuelan marines help the local fire brigade stop a potentially large waterfront fire at Fort-de-France, Martinique.[17]

Fortunately for the small British islands of the Lesser Antilles, no major city fires, such as the Port of Spain holocaust of March 1895, occurred in the first decade of the twentieth century after the British military withdrawal. But the Jamaica fire and earthquake of 1907 reminded everyone of the region's precarious physical geography and the ongoing need to seek outside help in the case of catastrophe. Nor was a major urban fire in the greater British Caribbean long in coming. On December 22, 1913, "Black Monday" saw an enormous fire destroy much of Georgetown, British Guiana. Traced to a Chinese fireworks shop, the fire raced through the city. Fueled by its own wooden buildings, it spread via showers of windblown sparks that ignited nearby and even distant roofs, until the fire brigade and volunteers brought it under control. The Georgetown fire, like major urban fires elsewhere in the region, led to substantial insurance claims against local and foreign companies whose policies protected local property, and it also inspired recollections of the similarly destructive Georgetown fires of earlier decades.[18]

Yet fire in the British Caribbean, while reinforcing the importance of past experiences, also was adapting itself to highlight the present and the future, just as it had thrown light on earlier decades. Early in 1906 a fire burned several buildings in King Street in Bridgetown, Barbados. It was caused from the explosion—of all things—of a "motor car." A local doctor had driven his vehicle north to St. James parish to attend to a patient. On the way back to town, the doctor noticed that the car was giving off heat, so he "readjusted the machinery." But the car exploded after midnight: "The doctor stated that the tank which held five gallons of oil when filled, contained about four gallons when he reached home that night."[19] Technology and political identities in the British West Indies eventually would change, but fire would always be there.

When it acted as a destructive agent, Caribbean fire at the turn of the century illuminated fundamental social and cultural characteristics of the region. Contested resources such as piped water or the usually uneven competition over varying uses of land always came into sharper, more immediate focus when fire was at hand. As an ecological force, fire played important roles, whether or not it was under human control. Kerosene fires brightened city streets after dark and—almost certainly far more important—extended waking hours by providing evening light for reading and conversation. Fire was a catalyst for and partner of the oppressed as they attempted to create better lives against high odds. Arson in city and countryside merited attention and notoriety. The human use of fire as a symbol of protest called attention to the social inequities that underpinned British colonialism. The torchlit *cannes brulées* processions, so meaningful in a historical sense, at once drew individuals together and kept local authorities rattled and off balance. Fires ignited in passion and anger over particular local events on individual islands brought regional problems to London's attention. Always in the background, sometimes at the forefront, and never under complete control, Caribbean fire invariably accompanied the people of the region as it always had in the past, and in some cases it lighted the way.

Notes

Abbreviations

CO British Colonial Office correspondence

PP British Sessional Papers (Parliamentary Papers)

RC *Report of the West India Royal Commission* (London: HMSO, 1897)

Preface

1. For a brief historical overview of newspapers in the Commonwealth Caribbean, see Lent, *Mass Communications in the Caribbean*, 1–16, and Pactor, *Colonial British Caribbean Newspapers*.

2. "The Growth of Lawlessness and Its Cause," *Barbados Agricultural Reporter*, March 1, 1901.

3. A history of the importance of reading newspapers as well as other print media would likely prove to be as valuable for students of the British Caribbean as it has been elsewhere. See Rose, *Intellectual Life of the British Working Classes*.

4. Denevan, "Bernard Q. Nietschmann."

Chapter 1

1. "New Troubles for the Planter," *Antigua Standard*, March 23, 1895.

2. "Incendiary Fires," *Port of Spain Gazette*, June 17, 1898.

3. The most influential American geographer of the twentieth century, the late Carl O. Sauer, wrote widely about the role of fire and human beings as ecological dominants on the surface of the earth, in the Caribbean region and elsewhere; see his "Man's Dominance by Use of Fire." Many other geographers have written about fire's influence, although none has been so influential as Sauer. See Vale, *Fire, Native Peoples, and the Natural Landscape*, for a very recent example.

4. Cronon, "Reading the Palimpsest," 358–59.

5. Professional responsibility and good manners suggest that citing this particular book here is inappropriate. Caribbeanists will surely know the text to which I refer.

6. The leading historian of fire, Stephen J. Pyne, has written widely about the subject. See, among his many articles and books, *Fire in America*, *World Fire*, and *Vestal Fire*. Also see Goudsblom, *Fire and Civilization*; Hazen and Hazen, *Keepers of the Flame*; and Rossotti, *Fire*. Books about the great London fire of 1666 continue to appear. For a very recent interpretation, see Hanson, *Great Fire of London*.

7. Salisbury, *Great Black Dragon Fire*, describes the massive fire in North China of

May 1987, one of the momentous environmental events of the late twentieth century that went almost unnoticed in the West. See also a recent book, also with a Eurasian setting but with an approach very similar to mine: Frierson, *All Russia Is Burning!*

8. Pyne, *Burning Bush*.

9. The use of "fire" in West Indian story, song, and general expression abounds. The best-known biography of one of the most famous West Indians in recent history is White, *Catch a Fire*. See also Smorkaloff, *If I Could Write This in Fire*.

10. See Grossman, *Political Ecology of Bananas*, for an example of a political ecological study within an eastern Caribbean context.

11. MacArthur and Wilson, *Theory of Island Biogeography*, 8–9.

12. Howard, "Vegetation of the Antilles."

13. Manning, "Nicknames and Number Plates in the British West Indies."

14. Richardson, *Economy and Environment*, 18–66. See also Beachey, *British West Indies Sugar Industry*.

15. "Balloon Ascent and Parachute Descent," *Port of Spain Gazette*, November 13, 1889.

16. "Popular Diversion," *Antigua Standard*, February 23, 1903.

17. *Antigua Standard*, January 16, 1884, and "The North American and West Indian Squadron," *Antigua Standard*, January 30, 1886.

18. "British Warships in the West Indies," *St. George's Chronicle and Grenada Gazette*, February 23, 1907.

19. Joseph, "Strategic Importance," 36.

20. "The U.S. Atlantic Fleet," *Voice of St. Lucia*, February 21, 1895.

21. "The Removal of the Troops from Barbados to St. Lucia," *Dominican*, March 28, 1895.

22. CO 884/8, no. 142, "Correspondence Relating to the Garrisons in the West Indies," September 1906, 10–13.

Chapter 2

1. CO 295/362, no. 89, "Fire at Port of Spain," March 7, 1895.

2. Langley, *United States and the Caribbean in the Twentieth Century*, 24–25.

3. "Great Fire in Haiti," *Barbados Globe and Colonial Advocate*, August 13, 1896.

4. Sir Ralph Williams, *How I Became a Governor*, 387–88.

5. Schwartz, "Hurricane of San Ciriaco," 334.

6. Garesche, *Complete Story of the Martinique and St. Vincent Horrors*, 118–40, 168.

7. "Hurricane Spreads Havoc in Jamaica," *New York Times*, August 13, 1903.

8. Cundall, *Historic Jamaica*, 154.

9. "The Earthquake: Extent of the Disaster," *The Times*, January 19, 1907.

10. CO 884/9, no. 164, "Correspondence Relating to the Landing of United States Naval Forces at Kingston, Jamaica, and the Resignation by Sir A. Swettenham of His Office as Governor of Jamaica," April 1907.

11. Tilchin, "Theodore Roosevelt, Anglo-American Relations, and the Jamaica Incident of 1907," 385, suggests that "it is time to bring this story out of the archives and into

the historical literature." See also Tilchin, *Theodore Roosevelt and the British Empire*, 117–68, on the Jamaica earthquake.

12. See, among many other sources, Langley, *United States and the Caribbean*; Healy, *Drive to Hegemony*; Collin, *Theodore Roosevelt's Caribbean*; Musicant, *Empire by Default*; and Traxel, *1898*.

13. Pérez, *Winds of Change*.

14. McCullough, *Path between the Seas*, 323–24. For the Mount Pelée catastrophe itself, see Zebrowski, *Last Days of St. Pierre*.

15. Richardson, "Catastrophes and Change on St. Vincent."

16. Ragatz, *Fall of the Planter Class*, 375–76.

17. Pattullo, *Fire from the Mountain*, 194.

18. Richardson, *Economy and Environment*, 89.

19. "The Fire of Monday," *Port of Spain Gazette*, March 6, 1895.

20. "The Hurricane in Our Sister Islands," *Port of Spain Gazette*, September 16, 1898.

21. "The Burning of Fort de France," *Voice of St. Lucia*, June 28, 1890.

22. "An Episode of the Saint Pierre Catastrophe," *Voice of St. Lucia*, June 5, 1902.

23. "The Condition of Haiti," *Port of Spain Gazette*, July 28, 1888.

24. Morison, *Admiral of the Ocean Sea*, 219.

25. Rouse, *Tainos*, 49–70.

26. Higuera-Gundy et al., "10,300 C-14 Year Record," 165.

27. Petersen, "Taino, Island Carib, and Prehistoric Amerindian Economies," 126.

28. Pyne, *Vestal Fire*, 413.

29. Sauer, *Early Spanish Main*, 59, 115, 244.

30. Rouse, *Tainos*, 16.

31. David Watts, *West Indies*, 55.

32. Petersen, "Taino, Island Carib, and Prehistoric Amerindian Economies," 120.

33. Murray, *Islands and the Sea*, 30.

34. Walker, *Columbus and the Golden World*, 160–61, 169.

35. Hulme, *Colonial Encounters*, 78.

36. Sauer, *Early Spanish Main*, 294. See also Crosby, *Columbian Exchange*, for a full discussion of the early ecological changes on Hispaniola.

37. Galloway, *Sugar Cane Industry*, 77–83.

38. Bridenbaugh and Bridenbaugh, *No Peace beyond the Line*, 42.

39. Quoted in Innes, "Pre-Sugar Era of European Settlement in Barbados," 3.

40. Kimber, *Martinique Revisited*, 109, 126.

41. Beard, *Natural Vegetation*.

42. Harris, *Plants, Animals, and Man in the Outer Leeward Islands*, 192–93.

43. Sheridan, *Doctors and Slaves*, 24–26. For the account of clearing Tobago, see Southey, *Chronological History*, 1:294–95.

44. Southey, *Chronological History*, 1:322–23.

45. David Watts, *West Indies*, 185–86.

46. See Kiple, *Caribbean Slave*, esp. 104–19.

47. Sheridan, *Doctors and Slaves*, 31–34.

48. Individual slave vessels seem to have sailed from West Africa to the Caribbean at various times of the year, depending on weather and agricultural conditions in both places. See Hugh Thomas, *Slave Trade*, 393–94.

49. Ibid., 396–97.

50. Handler and Lange, *Plantation Slavery in Barbados*, 54.

51. McDonald, *Economy and Material Culture of Slaves*, 101–2. On Barbados the combustibility of slave houses apparently inspired group efforts in fire fighting that transcended individual villages and plantations. See Handler, "Plantation Slave Settlements in Barbados," 145.

52. Oliver, *History of the Island of Antigua*, vol. 1, entries for August 1, 1786, and May 10, 1788.

53. Pares, *West-India Fortune*, 45.

54. Galloway, *Sugar Cane Industry*, 102.

55. Dirks, *Black Saturnalia*, 26.

56. Sheridan, *Sugar and Slavery*, 115.

57. Galloway, *Sugar Cane Industry*, 96–99.

58. Goveia, *Slave Society in the British Leeward Islands*, 164–65.

59. James, *Black Jacobins*, 88.

60. Gaspar, *Bondmen and Rebels*, 4–5.

61. Craton, *Testing the Chains*, 186–87, 261–62.

62. For an overview of some of the macrochanges in the region in this era, see Craton, "Transition from Slavery to Free Wage Labour in the Caribbean."

63. PP 1893–94/LX, "Annual Report for St. Lucia for 1891," 205.

64. Richardson, *Panama Money in Barbados*, 221.

65. Ramesar, "Patterns of Regional Settlement and Economic Activity," 199.

66. Normally "well-behaved" laborers, like the Barbadians who traveled for work in British Guiana, often were known to become antagonistic and surly on the job. See *RC*, Appendix C, "British Guiana," 79, which is available on British Sessional Paper microcards and in a more accessible form in vols. 7 and 8 of the bound British Parliamentary Papers published in 1971 by the Irish University Press.

67. Brereton, *Race Relations in Colonial Trinidad*, 110–15.

68. Lowenthal, *West Indian Societies*.

69. Brizan, *Grenada*, 214.

70. Baker, *Centring the Periphery*, 136.

71. *Dominican*, October 11, 1894.

72. Roberts, *West Indians and their Language*, 10, 13–14.

73. Brereton, *Race Relations in Colonial Trinidad*, 143.

74. Letter from Anglo-Catholicus, *Voice of St. Lucia*, March 24, 1888.

75. *RC*, Appendix C, pt. 12, "St. Kitts–Nevis," 235–36.

76. *Barbados Herald*, May 21, 1888.

77. Richardson, *Panama Money in Barbados*, 77–80.

78. Richardson, *Economy and Environment*, 109–10.

79. "The Return of the Royal Mail Steamers," *Barbados Agricultural Reporter*, April 13, 1903.

80. Richardson, "Human Mobility in the Windward Caribbean."

81. "The Carnival," *Port of Spain Gazette*, January 31, 1885.

82. *Dominican*, March 1, 1894.

83. Sir H. H. Bell, *Glimpses of a Governor's Life*, 24.

84. For detailed information about Port of Spain at the turn of the century, see Eversley, *Trinidad Reviewer*.

85. Magid, *Urban Nationalism*, 68.

86. U.S. National Archives, Despatches from U.S. Consuls in Trinidad, entry for July 10, 1887.

87. Pearse, "Carnival in Nineteenth Century Trinidad," 192.

88. Magid, *Urban Nationalism*, 71.

89. Eversley, *Trinidad Reviewer*, 99–100.

90. Green, *British Slave Emancipation*, 353.

91. "Death of a Political Veteran," *Dominica Guardian*, June 11, 1904.

92. *RC*, Appendix C, pt. 12, "Virgin Islands," 247.

93. "The Enforcement of Act 153," *Equilibrium*, September 24, 1885.

94. *RC*, Appendix C, pt. 12, "Antigua," 206.

95. *Dominican*, February 3, 1881, January 26, 1882, and May 15, 1890.

96. Laurence, "Trinidad Water Riot," 15.

97. Brereton, *Law, Justice, and Empire*, 287–88.

98. Richardson, "Detrimental Determinists."

99. J. J. Thomas, *Froudacity*, 132.

Chapter 3

1. "Queen Victoria's Bonfires," *Barbados Globe*, May 17, 1897.

2. "The Celebrations in the West Indies" *Voice of St. Lucia*, July 1, 1897.

3. "Soufrière," *Voice of St. Lucia*, July 8, 1897.

4. Rossotti, *Fire*, 146.

5. "The Fireworks," *Port of Spain Gazette*, June 24, 1887.

6. *Voice of St. Lucia*, August 11, 1888.

7. Dirks, *Black Saturnalia*, 167–68, describes precautions that West Indian planters took during Christmas revelries in the time of slavery.

8. Richardson, *Economy and Environment*, 61–62. Similarly, the disturbances described in O. Nigel Bolland's aptly titled *On the March: Labour Rebellions in the British Caribbean, 1934–39* underline how important torches were in the momentous anticolonial demonstrations of the 1930s.

9. Nicholls, "Legislation to Control Bush Fires," 87–88.

10. For a more detailed account of the Guy Fawkes resistance in Grenada, see Richardson, "'Respectable' Riot."

11. Brereton, *Race Relations in Colonial Trinidad*, is excellent on Port of Spain's social and cultural complexities at the time.

12. CO 884/4, no. 40, "Mr. Hamilton's Report on the Causes and Circumstances of the Disturbances in Connexion with the Carnival in Trinidad, 13 June 1881," 13.

13. "The Origin of Canne Boulee," *Port of Spain Gazette*, March 26, 1881.

14. CO 884/4, no. 40, "Mr. Hamilton's Report on the Causes and Circumstances of the Disturbances in Connexion with the Carnival in Trinidad, 13 June 1881," 5.

15. De Verteuil, *Years of Revolt*, 63–64.

16. Cowley, *Carnival*, 80.

17. Ibid., 78–79.

18. CO 884/4, no. 40, "Mr. Hamilton's Report on the Causes and Circumstances of the Disturbances in Connexion with the Carnival in Trinidad, 13 June 1881," 15–16.

19. "The Carnival," *Grenada People*, February 8, 15, 1894.

20. March 10, 1892; "Masquerading," February 23, 1893; February 8, 1894; and "A Timely Warning," February 21, 1895, all from *Dominican*.

21. Cowley, *Carnival*, 100.

22. "The Carnival Opens," *Port of Spain Gazette*, February 17, 1896.

23. "Port-of-Spain's Largest Fire," *Port of Spain Gazette*, March 7, 1895.

24. Carmichael, *History of Trinidad and Tobago*, 297; "Great Excitement in San Fernando," *Port of Spain Gazette*, March 6, 1895.

25. Some accounts even suggest that the cricketers played a role in fighting the fire and were "loudly applauded" by onlookers while doing so. See Besson, *Angostura Historical Digest*, 330.

26. CO 295/362, no. 89, "Fire at Port of Spain," March 7, 1895.

27. "Wrapped in Flames," *Port of Spain Gazette*, August 8, 1896.

28. CO 884/4, no. 46, "Fire at Kingston; despatch from the Governor," 1882.

29. "The Burning of Fort-de-France," *Port of Spain Gazette*, June 27, 1890.

30. "Fires at Colon and Martinique," April 27, 1896, and "Great Fire in Haiti," August 13, 1896, both from the *Barbados Globe and Colonial Advocate*.

31. "Fire in St. Andrew's," *Grenada People*, December 22, 1886.

32. "Fire at Grenville," *Grenada People*, December 22, 1886; CO 321/100, no. 126, "Grenville Water Supply," October 25 1887.

33. Letter from "A Liberal Catholic," *St. Christopher Advertiser and Weekly Intelligencer*, June 17, 1884.

34. *Independent*, October 30, 1884.

35. Hazen and Hazen, *Keepers of the Flame*, 39–40; Earnest, *Volunteer Fire Company*, 5–6.

36. "The Frederick Street Fire," *Port of Spain Gazette*, May 19, 1891.

37. "Fire!," *Dominican*, January 21, 1886.

38. CO 295/362, no. 89, "Fire at Port of Spain," March 7, 1895, 375, 377.

39. "Canefield Fires," *St. Christopher Advertiser and Weekly Intelligencer*, May, 18, 1897.

40. "St. Kitts," *Antigua Standard*, February 29, 1896.

41. "Fire at Basseterre," *St. Christopher Gazette and Caribbean Courier*, January 21, 1887.

42. *Barbados Globe and Colonial Advocate*, November 14, 1892.

43. "Destructive Fire in Castries," *Voice of St. Lucia*, August 10, 1899.

44. See November 16, 1893; "Roseau Volunteer Fire Brigade," June 23, 1898; and "Disastrous Fire in the Town of Portsmouth!," October 12, 1899, all from *Dominican*.

45. "Enquiry into the Late Fire," *Port of Spain Gazette*, March 18, 1895.

46. "Accidents from Kerosene Oil," *Barbados Herald*, May 17, 1886.

47. "Police Reports," *Barbados Herald*, November 10, 1881.

48. Yergin, *Prize*, 23.

49. *RC*, Appendix C, pt. 9, "Dominica," 134.

50. U.S. National Archives, Despatches from U.S. Consuls in Trinidad, entry for June 4, 1889.

51. Alleyne, *Historic Bridgetown*, 81.

52. *Dominican*, March 10, 1881.

53. *Equilibrium*, April 13, 1882.

54. "The Lighting of the Town," *Port of Spain Gazette*, October 27, 1891.

55. *Dominican*, September 7, 1882, and May 31, 1883.

56. "The Electric Light for St. John," *Antigua Standard*, May 18, 1895.

57. Yergin, *Prize*, 40.

58. Letter from "A Householder," *Dominican*, August 31, 1882.

59. *Dominican*, November 2, 1882.

60. "Accidents from Kerosene Oil," *Barbados Herald*, May 17, 1886.

61. "Dangerous Oil," *Port of Spain Gazette*, March 27, 1886.

62. *Port of Spain Gazette*, July 18, 1885.

63. Yergin, *Prize*, 50–51.

64. "Accidents from Kerosene Oil," *Barbados Herald*, May 17, 1886.

65. *Voice of St. Lucia*, March 7, 1907.

66. "The Burning of Fort de France," *Voice of St. Lucia*, June 28, 1890.

67. Carmichael, *History of Trinidad and Tobago*, 296.

68. Higgins, *History of Trinidad Oil*, 21, 40.

69. Richardson, *Economy and Environment*, 126–29.

70. Alleyne, *Historic Bridgetown*, 82–83.

71. "Lighting Castries," *Voice of St. Lucia*, December 17, 1892.

72. "The Electric Light," *Port of Spain Gazette*, October 17, 1894.

73. The plethora of electricity companies in Trinidad at the turn of the century was similar to the situation in London, where no fewer than sixty-five different companies provided electricity in the early 1900s. See Yergin, *Prize*, 79.

74. Gordon, *Century of West Indian Education*, 239.

75. Smith and Smith, *To Shoot Hard Labour*, 65.

76. *Dominica Guardian*, December 13, 1907.

77. *Equilibrium*, March 1, 1883.

78. Smith and Smith, *To Shoot Hard Labour*, 62.

79. Pyne, *Vestal Fire*, 47–48, on the invention and proliferation of the use of the "lucifer."

80. "The Red Star Match Factory," *Barbados Herald*, June 30, 1887; "The Match Factory," March 2, 1887, and "Fire at the Match Factory," October 20, 1888, both in *Port of Spain Gazette*.

81. "Fatal Accident at Toco," *Port of Spain Gazette*, September 4, 1889.

82. "Fire!," *Port of Spain Gazette*, October 26, 1898.

83. "Immorality and the Homes of Our Working Classes," *Port of Spain Gazette*, September 19, 1885.

84. "Fire in Chapman's Street," *Barbados Herald*, July 19, 1888.

85. *Voice of St. Lucia*, March 8, 1894.

86. *Barbados Official Gazette*, July 21, 1890.

87. "Fire!," *Antigua Standard*, August 16, 1890.

88. Trotman, *Crime in Trinidad*, 127.

89. *Barbados Herald*, April 12, 1888.

90. *Barbados Globe and Colonial Advocate*, March 16, 1891.

91. "The Recent Fire at Panama," *Dominican*, June 21, 1894.

92. CO 32/23, "The Report on the Fire Brigade for 1899," *Barbados Official Gazette*, January–June 1900, 375–76.

93. *St. Christopher Gazette and Caribbean Courier*, March 13, 1885.

94. *Grenada People*, November 3, 1892.

95. "Echoes of the Fortnight for Distant Readers," *Barbados Globe and Colonial Advocate*, October 8, 1891.

96. "The Late Fires," *Port of Spain Gazette*, September 29, 1896.

97. "Fire!," *Voice of St. Lucia*, April 14, 1898.

98. December 25, 1893; August 20, 1894; and "Fires at Trinidad," September 28, 1896, all in *Barbados Globe and Colonial Advocate*; "Incendiarism," *Barbados Agricultural Reporter*, March 2, 1898.

99. *Antigua Standard*, June 26, 1884.

100. "Incendiarism," *Dominican*, September 23, 1886.

101. "The Caroni Fire," *Port of Spain Gazette*, December 12, 1897.

102. "Cane Fire," *Barbados Globe and Colonial Advocate*, February 15, 1899.

Chapter 4

1. *Dominican*, September 18, 1880; "New Troubles for the Planter," *Antigua Standard*, March 23, 1895.

2. "On Bush Fires," *Dominica Guardian*, June 6, 1900; "Another Bush Fire," *Port of Spain Gazette*, May 26, 1899.

3. Francis Watts, "Tree Planting in Antigua," 408.

4. Nicholls, "Legislation to Control Bush Fires," 84.

5. Richardson, *Economy and Environment*, 172.

6. Pyne, *Vestal Fire*, 480.

7. Morris later became very influential in helping to forge tropical agricultural development policy throughout the British Caribbean and elsewhere in the British tropics. See Drayton, *Nature's Government*, 251–60.

8. Morris, *Forest Conservation in Jamaica*, 21–22, 32–33, 40–41.

9. CO 156/3, *Royal Gazette of the Leeward Islands*, August 19, 1882.

10. Fernow, *Brief History of Forestry*, 401.

11. For an overview of the development of international forestry in the late 1800s, see Pyne, *Vestal Fire*, 480–92.

12. Ibid., 480.

13. "Morris, Sir Daniel," *Colonial Office List for 1909*, 555; "Sir Daniel Morris," *St. Christopher Advertiser and Weekly Intelligencer*, February 23, 1909.

14. Beard, *Natural Vegetation*, 7.

15. *Dominican*, November 23, 1882.

16. Prestoe, *Report on Coffee Cultivation*, 4.

17. Hooper to the Under Secretary of State of the Colonial Office, March 14, 1888, in Morris, *Forest Conservation in Jamaica*, 122. All of E. D. M. Hooper's published reports on Caribbean forestry are in the archives of the Royal Botanic Gardens, Kew, England.

18. Ribbentrop, *Forestry in British India*, 148.

19. *Dominican*, May 24, 1888.

20. *Voice of St. Lucia*, June 12, 1886.

21. *Antigua Standard*, June 23, 1886.

22. Richardson, *Economy and Environment*, 53. From his post in the Windwards, Moloney moved on to become the governor of Trinidad and Tobago, where as discussed in Chapters 6 and 7, his handling of Port of Spain's water problems early in the 1900s was not nearly so successful.

23. "Willful Fire Setting," *Port of Spain Gazette*, April 20, 1894.

24. *Dominican*, May 17, 1888.

25. Brereton, *Race Relations in Colonial Trinidad*, 139.

26. Eversley, *Trinidad Reviewer*, 261.

27. "Young Man, Go East!," *Port of Spain Gazette*, August 14, 1889.

28. *Dominican*, February 3, 1881, and January 26, August 17, 1882; "St. Lucians, Beware!," *Dominica Guardian*, July 16, 1904.

29. *Voice of St. Lucia*, December 15, 1904; "Barbados and the Crown Lands of St. Lucia," *Voice of St. Lucia*, September 20, 1894.

30. Nicholls, "Legislation to Control Bush Fires," 31; Ballou, "Report by the Entomologist."

31. CO 156/3, *Royal Gazette of the Leeward Islands*, August 19, 1882.

32. Ribbentrop, *Forestry in British India*, 42.

33. "Lecture on Forestry," *Port of Spain Gazette*, March 13, 1903.

34. Letter from "X," *Antigua Standard*, January 30, 1889 (emphasis in original).

35. *RC*, 37.

36. CO 28/250, no. 200, "Forestry Report," September 13, 1899, 361.

37. "Mr. E. M. Hooper," *Voice of St. Lucia*, June 19, 1886; "The Evils of Deforestation," *St. George's Chronicle and Grenada Gazette*, June 29, 1895.

38. "The Deforestation of Our Valleys," *Port of Spain Gazette*, February 2, 1889.

39. Quilter, *Notes on a Visit*, 9.

40. "Conservation of Forests," *Grenada People*, November 17, 1887.

41. Beard, *Natural Vegetation*, 30. Some recent scholarship has extolled the King's Hill act as a glimmer of well-intentioned environmentalism in an era of market-driven ecological destruction. See Grove, "Origins of Western Environmentalism," as well as his better-known *Green Imperialism*.

42. "Our Forest Trees," *Port of Spain Gazette*, April 25, 1896.

43. "The Botanic Garden," *Grenada People,* November 17, 1886. The records of the early botanic stations in the eastern Caribbean are housed in bound volumes at the Royal Botanic Gardens, Kew, Surrey, England.

44. Prestoe, *Report on Coffee Cultivation,* 5–6.

45. Hart, *Trinidad,* 7.

46. Richardson, *Economy and Environment,* 220–21.

47. "The Louisiana Planter and Sugar Manufacturer," May 20, 1892, in *Windward Islands, Misc., 1878–98.* Bound volume in the Archives of the Royal Botanic Gardens, Kew, Surrey, England.

48. *Anguilla, St. Kitts–Nevis, Virgin Islands, 1874–1913,* 91. Bound volume in the Archives of the Royal Botanic Gardens, Kew, Surrey, England.

49. Cookman, *Virgin Islands,* 9.

50. *Barbados Herald,* February 19, 1885.

51. "Is Charcoal Edible?," *Antigua Standard,* November 23, 1901.

52. Hart, *Trinidad,* 14, 19.

53. Hooper, *Report upon the Forests of St. Vincent,* 9–10.

54. CO 884/5, no. 77, "Report of the Condition of St. Vincent," August 1897, 249.

55. CO 321/99, "Grievances of the Working Classes," May 7, 1899.

56. CO 884/5, no. 77, "Report of the Condition of St. Vincent," August 1897, 249.

57. *St. Vincent Witness,* November 4, 1880.

58. Hart, "Report on the Agri-Horticultural Resources of Tobago," 20.

59. "The Oropouche Lagoon on Fire," *Port of Spain Gazette,* June 15, 1889.

60. Nicholls, "Legislation to Control Bush Fires," 88–89.

61. "Echoes of the Fortnight for Distant Readers," *Barbados Globe and Colonial Advocate,* October 9, 1890. Of course the volcanic eruption in St. Vincent in 1902 was a massive "natural" fire in itself.

62. Hart, *Trinidad,* 4–5.

63. "Bush Fires," *Port of Spain Gazette,* May 29, 1898.

64. *Voice of St. Lucia,* May 9, 1901.

65. Francis Watts, "Tree Planting in Antigua," 409.

66. Nicholls, "Legislation to Control Bush Fires," 87.

67. DeBano, Neary, and Ffolliott, *Fire's Effects on Ecosystems,* 215.

68. Nicholls, "Legislation to Control Bush Fires," 86, 88.

69. Ibid., 89.

70. *Reports on the Botanic Station, Grenada,* 18.

71. "The Locusts," *Port of Spain Gazette,* December 26, 1885.

72. Nicholls, *Text-Book of Tropical Agriculture.*

73. *RC,* Appendix C, pt. 12, "St. Kitts–Nevis," 219.

74. "Legislative Council," *Dominican,* August 3, 1899.

75. "On Bush Fires," *Dominica Guardian,* June 6, 1900; "Harmfulness of Bush Fires," *Port of Spain Gazette,* February 23, 1901; Nicholls, "Legislation to Control Bush Fires."

76. Nicholls, "Legislation to Control Bush Fires," 81–82.

77. Ibid., 91–96.

78. "The Fire Ordinance," *Port of Spain Gazette,* May 7, 1899.

79. "Dr. Nicholls' Paper on 'Bush Fires,'" *Dominica Guardian*, June 13, 1900.

80. Ribbentrop, *Forestry in British India*, 154.

Chapter 5

1. See Wilkinson, *Big Sugar*, 21–28, for a vivid description of sugarcane fires.

2. Barnes, *Sugar Cane*, 345.

3. Knight, *Caribbean*, 179.

4. Barnes, *Sugar Cane*, 344.

5. Blackburn, *Sugar-cane*, 256.

6. See Davies, "Causes and Consequences of Cane Burning."

7. Professor Selwyn Carrington of Howard University offered an interesting disagreement with this point when I presented a variation of this chapter as a paper at the meeting of the Association of Caribbean Historians in the Bahamas in April 2002. He feels that there is a strong desire to destroy among those who start clandestine cane fires.

8. Barnes, *Sugar Cane*, 348.

9. *Dominican*, January 12, 1888.

10. *Dominican*, December 7, 1893.

11. See Deerr, *History of Sugar*, 1:193–203, for sugar production data for each island from the seventeenth to the twentieth century.

12. See chap. 8 in Richardson, *Economy and Environment*, for a discussion of how this perceived interisland homogeneity inspired "regional" land-use solutions for very different particular islands at the turn of the century.

13. Blackburn, *Sugar-cane*, 74–75.

14. *St. Vincent Witness*, June 10, 1880.

15. Richardson, *Economy and Environment*, 86–96.

16. *St. Kitts–Nevis Botanic Station, 1901–10*, 5. Bound volume at the archives of the Royal Botanic Garden, Kew, Surrey, England.

17. Haraksingh, "Labour, Technology, and the Sugar Estates."

18. *RC*, Appendix C, pt. 5, "Tobago," 361–62.

19. Johnson, "Origins and Early Development." Cane farming also was attempted on a limited scale at the time in Antigua and St. Lucia. See Carmody, "Cane-Farming in Trinidad," 40–41.

20. "West Indies," 522–23.

21. Richardson, *Caribbean Migrants*, 100–101.

22. *Independent*, December 20, 1883.

23. "Fire Insurance and Other Matters in Connection with Our Staple," *Antigua Standard*, March 3, 1894.

24. Galloway, "Botany in the Service of Empire"; García-Muñiz, "Interregional Transfer of Biological Technology in the Caribbean."

25. *International Sugar Journal* 3, no. 34 (October 1, 1901): 505–6.

26. Stubbs, *Sugar Cane*, 29; see 24–29 and 80–82 for a discussion of sugarcane's water content, a discussion remarkably similar to those in more modern texts.

27. Barnes, *Sugar Cane*, 31–32, 169–70.

28. Stubbs, *Sugar Cane*, 195–99.

29. Hagelberg, *Caribbean Sugar Industries*, 11.

30. Fawcett, "Prevention of the Introduction," 133.

31. "Canes by New Process," *Antigua Standard*, March 12, 1904.

32. Stubbs, *Sugar Cane*, 87–88.

33. Deerr, *Cane Sugar*, 122.

34. Richardson, *Economy and Environment*, 152–53.

35. *RC*, Appendix A, 134.

36. "The St. Lucia Agricultural Society," *Voice of St. Lucia*, July 25, 1907.

37. "Echoes of the Fortnight for Distant Readers," *Barbados Globe and Colonial Advocate*, February 25, 1892.

38. C. A. Barber, "Report on an Outbreak of Shot-Borer."

39. CO 321/161, "Report by Professor J. B. Harrison . . . ," April 11, 1895; CO 264/19, *St. Vincent Official Gazette*, May 10, 1894, 170.

40. "Valuable Evidence for the Defence," *Antigua Standard*, March 30, 1895.

41. *RC*, Appendix C, pt. 2, "British Guiana," 68.

42. Deerr, *Cane Sugar*, 122. A ratoon sugarcane crop grows directly from the cut stalks of the preceding season's canes with no replanting.

43. "Setting Fire," May 24, 1898, and "Fires in the Country," May 4, 1899, both in *Port of Spain Gazette*.

44. "Mr. Watts's Lectures," *St. Christopher Advertiser*, February 18, 1902.

45. "The Railway and Cane Fires," *Barbados Agricultural Reporter*, April 30, 1906.

46. Shephard, "Sugar Industry of the British West Indies," 166.

47. Barnes, *Sugar Cane*, 344, 355.

48. Letter from "A. Planter," *Antigua Standard*, October 30, 1889.

49. Barnes, *Sugar Cane*, 446.

50. Advertisement in the *Port of Spain Gazette*, February 11, 1906.

51. Barnes, *Sugar Cane*, 447.

52. *Independent*, June 12, 1884.

53. *Barbados Herald*, March 20, 1882.

54. *Barbados Globe and Colonial Advocate*, April 21, 1892.

55. "Fire in Tobago," *Port of Spain Gazette*, September 22, 1893.

56. "Fire!," *Dominican*, January 21, 1886.

57. *St. Vincent Witness*, July 14, 1881.

58. *St. Christopher Gazette and Caribbean Courier*, June 12, 1885.

59. "Setting Fire to Cane-trash," *Barbados Herald*, May 22, 1882.

60. *Antigua Standard*, June 26, 1884.

61. *Antigua Standard*, June 30, 1894.

62. *Grenada People*, November 3, 1892.

63. "Fire at Snug Corner," *St. George's Chronicle and Grenada Gazette*, August 3, 1895.

64. *St. George's Chronicle and Grenada Gazette*, July 28, 1897.

65. "The St. Kitts Rioters," *Barbados Globe and Colonial Advocate*, March 23, 1896.

66. "New Troubles for the Planter," *Antigua Standard*, March 23, 1895.

67. "Bridgetown: Thursday, April 18th, 1895," *Barbados Globe and Colonial Advocate*, April 18, 1895.

68. CO 28/254, no. 89, "Cane Fires," May 21, 1901, 410.

69. Letter from J. A. Cooper, *Voice of St. Lucia*, July 11, 1885.

70. *Independent*, November 11, 1886.

71. CO 28/254, no. 89, "Cane Fires," May 21, 1901, 412.

72. "Incendiarism," *St. Christopher Advertiser and Weekly Intelligencer*, February 19, 1901.

73. "Agriculture and Agricultural Labourers," *Barbados Agricultural Reporter*, February 12, 1901.

74. "Echoes of the Fortnight for Distant Readers," *Barbados Globe and Colonial Advocate*, January 2, 1890.

75. "The Fire Fiend," *St. Christopher Advertiser*, March 18, 1902.

76. "Echoes of the Fortnight for Distant Readers," *Barbados Globe and Colonial Advocate*, February 28, 1889.

77. "Destructive Fire in Castries," *Voice of St. Lucia*, August 10, 1899.

78. "Deputation to the Administrator," *Voice of St. Lucia*, August 17, 1899.

79. *St. Christopher Gazette and Caribbean Courier*, April 13, 1883.

80. "Incendiarism," *Antigua Standard*, March 25, 1905.

81. "Echoes of the Fortnight for Distant Readers," *Barbados Globe and Colonial Advocate*, February 12, 1891.

82. C. J. Manning, "The Volunteer Movement," *Barbados Agricultural Reporter*, January 31, 1900.

83. CO 28/255, no. 189, "Act 18 of 1901. Cane Fires (Prevention) Act," October 24, 1901, 304–8.

84. "Flogging for Arson," *Barbados Agricultural Reporter*, February 19, 1902.

85. "The Prevalence of Canefires," *Barbados Agricultural Reporter*, November 22, 1900.

86. Richardson, *Panama Money in Barbados*, 97–99.

87. "Agricultural Report and Packet Summary," *Barbados Agricultural Reporter*, November 24, 1900.

88. "New Troubles for the Planter," *Antigua Standard*, March 23, 1895; "Bridgetown: Thursday, February 21st, 1895," *Barbados Globe and Colonial Advocate*, February 21, 1895.

89. "Cane Fire," *Barbados Globe and Colonial Advocate*, February 15, 1899.

90. "How to Stop Cane Fires," *Barbados Agricultural Reporter*, January 26, 1900.

91. CO 28/254, no. 89, "Cane Fires," May 21, 1901, 411.

92. "The Caroni Fire," *Port of Spain Gazette*, December 12, 1897.

93. "One of the Recent Cedar Hill Fires," *Port of Spain Gazette*, March 9, 1893.

94. "Cane Fires in Barbados," *Barbados Agricultural Reporter*, April 16, 1901.

95. "Incendiary Fires," *Port of Spain Gazette*, June 17, 1898.

96. "Cane Fires," *Barbados Agricultural Reporter*, April 13, 1903.

97. "Report on the Police Force, 1902," *Barbados Agricultural Reporter*, April 28, 1902.

98. PP 1904/LVI, "Report for Barbados for 1902–3," 6–7.

Chapter 6

1. CO 295/362, no. 89, "Fire at Port of Spain," March 7, 1895.

2. CO 295/362, no. 213, "Fire at Port of Spain," March 21, 1895.

3. "The Fire," *Port of Spain Gazette*, February 24, 1891.

4. "Our Fire Brigade," *Port of Spain Gazette*, March 11, 1895.

5. CO 32/16, "Report of the Poor Law Inspector, July–December 1894," *Barbados Official Gazette*, 689.

6. *RC*, Appendix C, pt. 11, "Antigua," 197.

7. "The Drought in Jamaica," *Port of Spain Gazette*, March 26, 1897.

8. Hardy, "Water and the Search for Public Health in London," 271.

9. Koeppel, *Water for Gotham*.

10. Hazen and Hazen, *Keepers of the Flame*, 129.

11. Ayers, *Promise of the New South*, 74.

12. Eversley, *Trinidad Reviewer*, 98.

13. Alleyne, *Historic Bridgetown*, 81.

14. "Gros-Islet, St. Lucia," *Voice of St. Lucia*, May 16, 1901.

15. British army detachments' experiences with disease incidence in the region in the late 1800s involved experimenting with and innovating methods of supplying fresh water, developments they shared with local civilian administrators. See Curtin, *Death by Migration*.

16. "The Situation," *Barbados Agricultural Reporter*, October 5, 1894. See Richardson, *Panama Money in Barbados*, 72–80, for a discussion of disease in Barbados at the turn of the century.

17. Green, *British Slave Emancipation*, 312–13.

18. *Dominican*, December 5, 1889.

19. "The Public Health," *Port of Spain Gazette*, February 11, 1893.

20. U.S. National Archives, Despatches from U.S. Consuls in St. Christopher, entry for July 20, 1903.

21. Richardson, *Economy and Environment*, 106, 108.

22. CO 152/154, no. 213, "Sandy Point and Dieppe Bay Water Works," July 20, 1883.

23. CO 321/97, no. 127, "Kingstown Water Supply," October 3, 1886.

24. "Echoes of the Fortnight for Distant Readers," *Barbados Globe and Colonial Advocate*, April 10, 1890.

25. *Voice of St. Lucia*, March 25, 1897.

26. "Water! Water Everywhere!," *Voice of St. Lucia*, February 21, 1885.

27. PP 1898/LIX, "Report for Barbados for 1897," 304.

28. CO 152/153, no. 163, "Death Rate in the District of Sandy Point & Dieppe Bay Water Works."

29. *Barbados Herald*, September 6, 1883.

30. *Voice of St. Lucia*, February 17, 1898.

31. *Dominican*, March 23, April 6, 1882, and March 2, 1893.

32. *Dominican*, July 17, 1880.

33. "Legislative Assembly," *Dominican*, June 11, 1885.

34. *Antigua Standard*, August 2, 1890.

35. "Carelessness at the Reservoir," *Equilibrium*, June 3, 1886 (emphasis in original).

36. CO 321/86, no. 27, "Suspension of Mr. Risk," March 1, 1885.

37. Untitled excerpt, July 14, 1898, and "The Insolence of Jack in Office," *Voice of St. Lucia*, March 2, 1899.

38. Ball, *Life's Matrix*, 222.

39. "The Deforestation of Our Valleys," *Port of Spain Gazette*, February 2, 1889.

40. "Water Conservancy," *Port of Spain Gazette*, March 14, 1885.

41. *Dominican*, May 24, 1888.

42. CO 152/154, no. 213, "Sandy Point and Dieppe Bay Water Works," July 20, 1883.

43. Levy, *Emancipation, Sugar, and Federalism*, 126–27.

44. "Big Fire at Carriacou," *Federalist and Grenada People*, July 18, 1904.

45. *Antigua Standard*, October 7, 1893.

46. *Equilibrium*, October 3, 1882.

47. "Barbados," *Dominican*, January 31, 1880.

48. *Barbados Herald*, December 25, 1882.

49. *Barbados Herald*, June 2, 1884.

50. *Voice of St. Lucia*, November 18, 1893.

51. "Trial of the New Steam Fire Engine for Castries," *Voice of St. Lucia*, November 25, 1893.

52. Oliver, *History of the Island of Antigua*, vol. 1, entry for May 10, 1788.

53. Higman, *Slave Populations of the British Caribbean*, 256.

54. *Dominican*, August 2, 1883.

55. *Grenada People*, May 17, 1888.

56. *Voice of St. Lucia*, April 21, 1898.

57. *Dominican*, October 9, 1890.

58. "Roseau Volunteer Fire Brigade," *Dominican*, June 23, 1898.

59. Green, *British Slave Emancipation*, 393.

60. "Mr. Hamilton's Report . . . Carnival in Trinidad," *Port of Spain Gazette*, October 29, 1881.

61. Trotman, *Crime in Trinidad*, 99.

62. Eversley, *Trinidad Reviewer*, 143.

63. "Monday Morning's Fire," *Port of Spain Gazette*, August 23, 1892.

64. "Report on the Fire Brigade," *Port of Spain Gazette*, November 22, 1893.

65. *Port of Spain Gazette*, October 26, 1887.

66. *Voice of St. Lucia*, December 13, 1894.

67. Earnest, *Volunteer Fire Company*, 144. See also Hazen and Hazen, *Keepers of the Flame*, 136–49, on "The Fireman Mystique." For a post–September 11 example, see Golway, *So Others Might Live*.

68. Asbury, *Ye Olde Fire Laddies*.

69. "The Port of Spain Fire Brigade," *Port of Spain Gazette*, September 7, 1900.

70. "The Fire Engine," March 30, 1896, and "Abundance of Water Supply: The New Fire Engine Proved This," April 11, 1896, *Port of Spain Gazette*.

71. Hazen and Hazen, *Keepers of the Flame*, 122–23.

72. CO 32/23, "The Report on the Fire Brigade for 1899," *Barbados Official Gazette*, January–June 1900, 375–77.

73. "The 'New Theory' in Practice and Its Pretensions," August 25, 1896, and advertisement, November 7, 1896, *Port of Spain Gazette*.

74. "Great Fire at Codrington College," *Antigua Standard*, May 6, 1885.

75. *St. Christopher Advertiser and Weekly Intelligencer*, February 7, 1888.

76. Editorial, *Barbados Herald*, March 22, 1883.

77. Cockerell and Green, *British Insurance Business*, 26–28.

78. Oliver, *History of the Island of Antigua*, vol. 1, entry for May 10, 1788.

79. CO 295/362, no. 89, "Fire at Port of Spain," March 7, 1895, 383.

80. "Local Fire Insurance Company," *Port of Spain Gazette*, May 22, 1886.

81. Karch with Carter, *Rise of the Phoenix*.

82. "Trinidad Fire Insurance Company," May 1, 1904; "The Local Insurance Company," June 26, 1904; "The Local Insurance Company," October 20, 1904; and "The Trinidad Fire Insurance Company Limited," December 13, 1904, all in *Port of Spain Gazette*.

83. *St. Vincent Witness*, July 14, 1881.

84. "Incendiary Fires," *Port of Spain Gazette*, January 17, 1885.

85. "Fires at Trinidad," *Barbados Globe and Colonial Advocate*, September 28, 1896.

86. Hazen and Hazen, *Keepers of the Flame*, 134.

87. "The Fort-de-France Fire," *Port of Spain Gazette*, July 1, 1890.

88. "A Chapter of West Indian History," *Antigua Standard*, January 19, 1907.

89. "The Fire Brigade," *Dominica Guardian*, June 27, 1903.

90. Magid, *Urban Nationalism*, 109.

91. "About Fire Insurance," *Port of Spain Gazette*, February 28, 1896.

92. "The Racine and the Insurance Companies," *Port of Spain Gazette*, November 28, 1897.

93. "The Fire Brigade and the Insurance Companies," *Port of Spain Gazette*, April 14, 1899.

94. "Fire Insurance Rates," *Port of Spain Gazette*, January 8, 1902.

95. "Water Supply and Fire Insurance," *Port of Spain Gazette*, May 28, 1902.

96. See Magid, *Urban Nationalism*, 141–44, for a discussion of the possible water supplies for Port of Spain in the 1890s.

97. Eversley, *Trinidad Reviewer*, 111–12, 265–70.

98. "A Public Works Difficulty," *Port of Spain Gazette*, February 5, 1901.

99. Eversley, *Trinidad Reviewer*, 112.

100. Hart, *Trinidad*, 4–5.

101. "The Filthy State of the Maraval Reservoir," *Port of Spain Gazette*, June 29, 1889.

102. CO 295/374, no. 301, "Port of Spain Water & Sewerage Works," 4.

103. "Board of Health," *Port of Spain Gazette*, December 4, 1895. See also Eversley, *Trinidad Reviewer*, 112.

104. "The Water Crisis," *Port of Spain Gazette*, January 19, 1889.

105. "The Voice of the People Is Heard," *Port of Spain Gazette*, August 25, 1896.

106. CO 295/374, no. 301, "Port of Spain Water & Sewerage Works."

107. Magid, *Urban Nationalism*, 107.

108. Ibid., 111.

109. "The Great Fire," *Port of Spain Gazette*, March 30, 1902.

110. "The Water Scandal," *Port of Spain Gazette*, April 2, 1902.

111. "Water Supply of Port-of-Spain," *Federalist and Grenada People*, May 1, 1902.

112. Magid, *Urban Nationalism*, 118; "The Water Question: Public Meeting at the Princes Building," *Port of Spain Gazette*, October 18, 1902.

113. PP 1903/XLIV, "Papers Relating to the Recent Disturbances at Port of Spain, Trinidad," 677.

114. Magid, *Urban Nationalism*, 122.

115. "Why Pay the Water Rates?," *Port of Spain Gazette*, February 28, 1903.

116. PP 1903/XLIV, "Report of the Commission of Enquiry into the Recent Disturbances at Port of Spain, Trinidad," July 1903, 710.

Chapter 7

1. CO 264/17, *St. Vincent Official Gazette*, September 25, 1890.

2. "Fire at Colon," *Antigua Standard*, September 27, 1890.

3. "The St. Kitts Riots," *Barbados Globe and Colonial Advocate*, February 24, 1896.

4. *Voice of St. Lucia*, March 26, 1903.

5. CO 28/221, "Act 20 of 1886–7," August 9, 1886.

6. Heuman, *"Killing Time"*; Levy, *Emancipation, Sugar, and Federalism*.

7. *Dominican*, July 27, 1882.

8. Le Bon, *Crowd*, 36.

9. Rudé, *Crowd in History*, 238, 241.

10. McClelland, *Crowd and the Mob*, 196.

11. Hobsbawm, *Labouring Men*.

12. CO 321/108, "Arrival of JaJa," June 9, 1888. See a more extended description of JaJa's arrival in St. Vincent in Richardson, *Economy and Environment*, 97–99.

13. *Voice of St. Lucia*, July 28, 1888.

14. *RC*, Appendix C, pt. 3, "Barbados," 221.

15. Brereton, *Law, Justice, and Empire*, 297–98.

16. "Sir John Gorrie in Tobago," *Port of Spain Gazette*, January 30, 1889.

17. "The West Indies," *The Times*, September 12, 1899. Fear of Haiti-like rebellion accented with fire was similarly widespread in the United States during the years of the revolt in French St. Domingue. See Berlin, *Many Thousands Gone*, 361–62.

18. Froude, *English in the West Indies*.

19. "Another Revolution in Hayti," *Dominican*, August 16, 1888.

20. "Burn St. Pierre," *Dominican*, May 7, 1891.

21. "Serious Riots at Martinique," *Voice of St. Lucia*, February 15, 1900.

22. *Equilibrium*, November 8, 1883, and November 13, 1884.

23. "Prospective Trouble," *Port of Spain Gazette*, December 10, 1887.

24. *Dominican*, March 11, 1897.

25. Ramdin, *Arising from Bondage*, 89–90.

26. De Verteuil, *Years of Revolt*, 128, 135.

27. Haraksingh, "Control and Resistance among Indian Workers," 75.

28. *RC*, Appendix C, pt. 4, "Trinidad," 344 (emphasis in original).

29. *Voice of St. Lucia*, December 28, 1905.

30. CO 884/9, no. 147, "Notes on West Indian Riots, 1881–1903," March 1905, 5–6; Boromé, "How Crown Colony Government Came to Dominica," 40.

31. *RC*, Appendix C, pt. 11, "Antigua," 206.

32. H. J. Bell, *Obeah*, 30.

33. L. Moore, Commercial Agency of the United States, St. Christopher, February 24, 1896, to Edward Uhl, Assistant Secretary of State, U.S. National Archives, Despatches from U.S. Consuls in St. Christopher.

34. Richards, "Order and Disorder," 6. The most complete description of the 1896 St. Kitts disturbances may be found in Richards's Ph.D. dissertation, "Masters and Servants."

35. Richards, "Order and Disorder," 7.

36. Ibid., 8.

37. *St. Christopher Advertiser and Weekly Intelligencer*, March 2, 1896; Richards, "Collective Violence in Plantation Societies."

38. "St. Kitts," *Antigua Standard*, February 29, 1896.

39. "Riot in St. Kitts," *Antigua Standard*, February 22, 1896.

40. "St. Kitt's under Martial Law," *New York Times*, March 13, 1896.

41. Richards, "Masters and Servants," 173.

42. "Negro Rebellion in St. Kitts," *New York Times*, February 29, 1896.

43. I argue that the riots inspired the 1897 commission, in Richardson, "Depression Riots and the Calling of the 1897 West India Royal Commission."

44. *RC*, Appendix C, pt. 1, "London," 187. In the eyes of the Colonial Office, all of the West Indian riots were grouped together. See CO 884/9, no. 147, "Notes on West Indian Riots, 1881–1903," March 1905, 1–16.

45. CO 884/9, no. 147, "Notes on West Indian Riots, 1881–1903," 14.

46. General descriptions of the riot exist in several places. See Magid, *Urban Nationalism*, 121–30; Eric Williams, *History of the People of Trinidad and Tobago*, 181–86; Laurence, "Trinidad Water Riot."

47. Whether or not Brake took proper precautions by securing enough police is debatable. See Laurence, "Trinidad Water Riot," 12.

48. PP 1903/XLIV, "Papers Relating to the Recent Disturbances at Port of Spain, Trinidad," 659.

49. "The Red House Stoned," *Port of Spain Gazette*, March 27, 1903.

50. PP 1903/XLIV, "Papers Relating to the Recent Disturbances at Port of Spain, Trinidad," 637.

51. PP 1903/XLIV, "Report of the Commission of Enquiry into the Recent Disturbances at Port of Spain, Trinidad," July 1903, 701.

52. PP 1903/XLIV, "Papers Relating to the Recent Disturbances at Port of Spain, Trinidad," 683.

53. "The Red House Stoned," *Port of Spain Gazette*, March 27, 1903.

54. PP 1903/XLIV, "Papers Relating to the Recent Disturbances at Port of Spain, Trinidad," 654–57.

55. PP 1903/XLIV, "Report of the Commission of Enquiry into the Recent Disturbances at Port of Spain, Trinidad," July 1903, 723.

56. CO 884/7, no. 122, "Correspondence Relating to Affairs in Trinidad," February 1904, 42.

57. "The Trinidad Riot," *Barbados Agricultural Reporter*, March 25, 1903.

58. *Voice of St. Lucia*, March 26, 1903.

59. "The Trinidad Riot," *The Times*, March 26, 1903; "Trinidad's Governor Blamed," *New York Times*, March 26, 1903.

60. PP 1903/XLIV, "Report of the Commission of Enquiry into the Recent Disturbances at Port of Spain, Trinidad," July 1903, 693–724.

61. PP 1904/LX, "Further Papers Relating to the Disturbances at Port of Spain, Trinidad, in March, 1903," April 1904, 409–31.

62. PP 1903/XLIV, "Papers Relating to the Recent Disturbances at Port of Spain, Trinidad," 674.

63. PP 1903/XLIV, "Report of the Commission of Enquiry into the Recent Disturbances at Port of Spain, Trinidad," July 1903, 705.

64. PP 1903/XLIV, "Papers Relating to the Recent Disturbances at Port of Spain, Trinidad," 638 (emphasis added).

65. "The Red House Stoned," *Port of Spain Gazette*, March 27, 1903.

66. PP 1903/XLIV, "Report of the Commission of Enquiry into the Recent Disturbances at Port of Spain, Trinidad," July 1903, 703.

67. PP 1904/LX, "Further Papers Relating to the Disturbances at Port of Spain, Trinidad, in March, 1903," April 1904, 418.

68. PP 1903/XLIV, "Report of the Commission of Enquiry into the Recent Disturbances at Port of Spain, Trinidad," July 1903, 701.

69. Ibid., 704.

70. CO 884/7, no. 122, "Correspondence Relating to Affairs in Trinidad," February 1904, 126–27.

71. PP 1903/XLIV, "Papers Relating to the Recent Disturbances at Port of Spain, Trinidad," 654–57.

72. Brereton, *Race Relations in Colonial Trinidad*, 110.

73. "Mr. Chamberlain on the Riot," *Port of Spain Gazette*, February 6, 1904.

74. *Voice of St. Lucia*, April 27, 1907.

75. "Strikes and Riots," *Voice of St. Lucia*, April 27, 1907. Unless indicated otherwise, the chronology of the April 1907 disturbances in St. Lucia is taken from this newspaper article.

76. Sir Ralph Williams, *How I Became a Governor*, 383.

77. Ibid., 384.

78. CO 321/235, no. 136, "Riots," April 30, 1907.

79. "Strikes and Riots," *Voice of St. Lucia*, April 27, 1907.

80. "Disturbances in St. Lucia," April 25, 1907; "The St. Lucia Riots," April 27, 1907; and "The Riots in St. Lucia," April 29, 1907, all in *The Times*.

81. "Big Meeting at Castries: For the Formation of a Volunteer Corps," *Voice of St. Lucia*, May 9, 1907.

82. Sir Ralph Williams, *How I Became a Governor*, 391.

83. Ibid., 387–88.

84. Richardson, *Economy and Environment*, 39–42.

85. Sir Ralph Williams, *How I Became a Governor*, 386.

86. CO 884/8, no. 142, "Correspondence Relating to the Garrisons in the West Indies," September 1906, 27–30.

Chapter 8

1. *Voice of St. Lucia*, April 27, 1907.

2. Joseph, "Strategic Importance," 42.

3. CO 884/8, no. 142, "Correspondence Relating to the Garrisons in the West Indies," September 1906, 6.

4. Ibid., 70–71.

5. "The Removal of the Troops from the West Indies," *Barbados Agricultural Reporter*, October 28, 1905.

6. Will, "Colonial Policy and Economic Development," 143.

7. "More Troops for Trinidad," *Barbados Agricultural Reporter*, April 28, 1903.

8. Henry N. Hall, "The Value of Human Life in the West Indies," *Federalist and Grenada People*, September 12, 1903.

9. Laurence, "Trinidad Water Riot," 15.

10. Scott, "Through the Diameter of Respectability."

11. Joseph, "Strategic Importance," 42.

12. Sir Ralph Williams, *How I Became a Governor*, 390.

13. "The Earthquake: An Eyewitness's Narrative," *The Times*, February 2, 1907.

14. "Drift," *Antigua Standard*, March 23, 1907.

15. "The Americans and Jamaica," *Federalist and Grenada People*, January 31, 1907.

16. "Fire at Grenville," *Federalist and Grenada People*, March 18, 1905.

17. "Our French Neighbours," *Dominica Guardian*, September 1, 1905.

18. *Black Monday*.

19. "Fire in King Street, Barbados: Caused by a Motor Car," *Port of Spain Gazette*, January 12, 1906.

Bibliography

Manuscript Collections

British Colonial Office correspondence; originals at the London Public Record Office and some items on microfilm at the Institute of Commonwealth Studies at the University of London

British Sessional Papers (Parliamentary Papers), published on microcards

Foreign and Commonwealth Office, London

Herbarium and Royal Botanic Gardens Library at Kew, Surrey, England

U.S. National Archives, Despatches from U.S. Consuls in Barbados, St. Christopher, and Trinidad (microfilm)

Newspapers

Antigua Standard

Antigua Times

Barbados Agricultural Reporter

Barbados Globe and Colonial Advocate

Barbados Herald

Dominica Guardian

Dominican

Equilibrium (Grenada)

Federalist (Grenada)

Federalist and Grenada People

Grenada People

Independent (St. Kitts)

Mirror (Trinidad)

New York Times

Port of Spain Gazette (Trinidad)

St. Christopher Advertiser (St. Kitts)

St. Christopher Advertiser and Weekly Intelligencer (St. Kitts)

St. Christopher Gazette and Caribbean Courier (St. Kitts)

St. George's Chronicle and Grenada Gazette

St. Kitts Commercial News

St. Vincent Witness

The Times (London)

Voice of St. Lucia

Books, Essays, Articles, and Dissertations

Alleyne, Warren. *Historic Bridgetown*. [Bridgetown]: Barbados National Trust, 1978.

Asbury, Herbert. *Ye Olde Fire Laddies*. New York: Knopf, 1930.

Ayers, Edward L. *The Promise of the New South: Life after Reconstruction*. New York: Oxford University Press, 1992.

Aykroyd, W. R. *Sweet Malefactor: Sugar, Slavery, and Human Society*. London: Heinemann, 1967.

Baker, Patrick L. *Centring the Periphery: Chaos, Order, and the Ethnohistory of Dominica*. Montreal: McGill-Queen's University Press, 1994.

Ball, Philip. *Life's Matrix: A Biography of Water*. New York: Farrar, Straus, and Giroux, 1999.

Ballou, H. A. "Report by the Entomologist on a Visit to the Northern Islands." In *West Indies Imp. Comm. Agric. 1917–27*. Bound volume in the archives of the Royal Botanic Gardens, Kew, Surrey, England, 1919.

Barber, C. A. "Report on an Outbreak of Shot-Borer in St. Kitts Sugar Estates." Circular no. 2. Supplement to the *Leeward Islands Gazette*, 1893.

Barnes, A. C. *The Sugar Cane*. New York: John Wiley and Sons, 1974.

Beachey, R. W. *The British West Indies Sugar Industry in the Late Nineteenth Century*. 1957. Reprint, Westport, Conn.: Greenwood Press, 1978.

Beard, J. S. *The Natural Vegetation of the Windward and Leeward Islands*. Oxford Forestry Memoirs 21. Oxford: Clarendon Press, 1949.

Beckwith, Martha W. *Black Roadways: A Study of Jamaican Folk Life*. Chapel Hill: University of North Carolina Press, 1929.

Bell, Sir H. H. *Glimpses of a Governor's Life from Diaries, Letters, and Memoranda*. London: Sampson, Low and Marston, 1946.

Bell, H. J. *Obeah: Witchcraft in the West Indies*. 1889. Reprint, Westport, Conn.: Negro Universities Press, 1970.

Berlin, Ira. *Many Thousands Gone: The First Two Centuries of Slavery in North America*. Cambridge: Harvard University Press, 1998.

Besson, Gerard A. *The Angostura Historical Digest of Trinidad and Tobago*. Trinidad: Paria, 2001.

Blackburn, Frank. *Sugar-cane*. New York: Longman, 1984.

Black Monday: Destructive Fire in Georgetown, December 22nd, 1913. Reprinted from the *Daily Chronicle*. Demerara, British Guiana: C. K. Jardine, 1914.

Blake, Nelson Manfred. *Water for the Cities: A History of the Urban Water Supply Problem in the United States*. Syracuse: Syracuse University Press, 1956.

Blaut, James M. "The Ecology of Tropical Farming Systems." In *Plantation Systems of the New World*, 83–103. Washington, D.C.: Pan American Union, 1959.

Bolland, O. Nigel. *On the March: Labour Rebellions in the British Caribbean, 1934–39*. Kingston, Jamaica: Ian Randle, 1995.

Boromé, Joseph A. "How Crown Colony Government Came to Dominica by 1898." *Caribbean Studies* 9 (1969): 26–67.

Boucher, Philip P. *Cannibal Encounters: Europeans and Island Caribs, 1492–1763*. Baltimore: Johns Hopkins University Press, 1992.

Bovell, John R. "Hints on the Planting and Cultivation of the Sugar-Cane and Intermediate Crops." In *Lectures to Sugar Planters*, edited by the Imperial Commissioner of Agriculture for the West Indies, 93–122. London: Dulau and Co., 1906.

Brereton, Bridget. *Law, Justice, and Empire: The Colonial Career of John Gorrie, 1829–1892*. Kingston, Jamaica: Press University of the West Indies, 1997.

———. *Race Relations in Colonial Trinidad, 1870–1900*. Cambridge: Cambridge University Press, 1979.

Bridenbaugh, Carl, and Roberta Bridenbaugh. *No Peace beyond the Line: The English in the Caribbean, 1624–1690*. New York: Oxford University Press, 1972.

Brizan, George. *Grenada, Island of Conflict: From Amerindians to People's Revolution*. London: Zed, 1984.

Carmichael, Gertrude. *The History of the West Indian Islands of Trinidad and Tobago*. London: Alvin Redman, 1961.

Carmody, P. "Cane-Farming in Trinidad." *West Indian Bulletin* 2 (1904): 33–41.

Chamoiseau, Patrick. *Texaco*. Translated by Rose-Myriam Rejouis and Val Vinokurov. New York: Pantheon, 1997.

Cockerell, H. A. L., and Edwin Green. *The British Insurance Business: A Guide to Its History and Records*. 2nd ed. Sheffield: Sheffield Academic Press, 1994.

Collin, Richard H. *Theodore Roosevelt's Caribbean: The Panama Canal, the Monroe Doctrine, and the Latin American Context*. Baton Rouge: Louisiana State University Press, 1990.

The Colonial Office List for 1909. London: Waterlow and Sons, 1909.

Cookman, N. G. *Virgin Islands: Report on the Condition of the Islands during 1897*. London: HMSO, 1898.

Cowley, John. *Carnival, Canboulay, and Calypso: Traditions in the Making*. Cambridge: Cambridge University Press, 1996.

Craton, Michael. *Testing the Chains: Resistance to Slavery in the British West Indies*. Ithaca: Cornell University Press, 1982.

———. "The Transition from Slavery to Free Wage Labour in the Caribbean, 1780–1890: A Survey with Particular Reference to Recent Scholarship." *Slavery and Abolition* 13, no. 2 (1992): 37–67.

Cronon, William. "Reading the Palimpsest." In *Discovering the Chesapeake: The History of an Ecosystem*, edited by Philip D. Curtin, Grace S. Brush, and George W. Fisher, 355–73. Baltimore: Johns Hopkins University Press, 2001.

Crosby, Alfred W. *The Columbian Exchange: Biological and Cultural Consequences of 1492*. Westport, Conn.: Greenwood Press, 1972.

Crowley, Daniel J. "Festivals of the Calendar in St. Lucia." *Caribbean Quarterly* 4, no. 2 (1955): 99–121.

Cundall, Frank. *Historic Jamaica*. London: Published for the Institute of Jamaica by the West India Committee, 1915.

Curtin, Philip D. *Death by Migration: Europe's Encounter with the Tropical World in the Nineteenth Century*. Cambridge: Cambridge University Press, 1989.

Davies, John. "The Causes and Consequences of Cane Burning in Fiji's Sugar Belt." *Journal of Pacific Studies* 22 (1998): 1–25.

Davis, Mike. *Late Victorian Holocausts: El Niño Famines and the Making of the Third World*. New York: Verso, 2001.

DeBano, Leonard F., Daniel G. Neary, and Peter F. Ffolliott. *Fire's Effects on Ecosystems*. New York: John Wiley, 1998.

Deerr, Noel. *Cane Sugar: A Text-Book on the Agriculture of the Sugar Cane, the Manufacture of Cane Sugar, and the Analysis of Sugar House Products*. Manchester: Norman Rodger, 1911.

——. *The History of Sugar*. 2 vols. London: Chapman and Hall, 1949–50.

Denevan, William M. "Bernard Q. Nietschmann, 1941–2000." *Geographical Review* 92, no. 1 (2002): 104–9.

de Verteuil, Fr. Anthony. *The Years of Revolt: Trinidad, 1881–1888*. Port of Spain: Paria, 1984.

Dirks, Robert. *The Black Saturnalia: Conflict and Its Ritual Expression on British West Indian Slave Plantations*. Gainesville: University of Florida Press, 1987.

Dobbin, Jay D. *The Jombee Dance of Montserrat: A Study of Trance Ritual in the West Indies*. Columbus: Ohio State University Press, 1986.

Drayton, Richard. *Nature's Government: Science, Imperial Britain, and the "Improvement" of the World*. New Haven: Yale University Press, 2000.

Dye, Alan. *Cuban Sugar in the Age of Mass Production: Technology and the Economics of the Sugar Central*. Stanford: Stanford University Press, 1998.

Earnest, Ernest. *The Volunteer Fire Company: Past and Present*. New York: Stein and Day, 1979.

Elder, J. D. "Color, Music, and Conflict: A Study of Aggression in Trinidad with Reference to the Role of Traditional Music." *Ethnomusicology* 8, no.2 (1964): 128–36.

Eversley, T. Fitz-Evan, comp. *The Trinidad Reviewer for the Year 1900*. London: Robinson, 1900.

Fawcett, William. "The Prevention of the Introduction and Spread of Fungoid and Insect Pests in the West Indies." *West Indian Bulletin* 1 (1900): 133–37.

Fernow, Bernard E. *A Brief History of Forestry: In Europe, the United States, and Other Countries*. 3rd rev. ed. Toronto: University Press, 1913.

Frazer, James George. *The Golden Bough: A Study in Magic and Religion*. New York: Macmillan, 1951.

Frierson, Cathy A. *All Russia Is Burning! A Cultural History of Fire and Arson in Late Imperial Russia*. Seattle: University of Washington Press, 2002.

Froude, James Anthony. *The English in the West Indies*. London: Longmans, Green, 1888.

Galloway, J. H. "Botany in the Service of Empire: The Barbados Cane-Breeding Program and the Revival of the Caribbean Sugar Industry, 1880s–1930s." *Annals of the Association of American Geographers* 86, no. 4 (1996): 682–706.

——. *The Sugar Cane Industry: An Historical Geography from Its Origins to 1914*. Cambridge: Cambridge University Press, 1989.

García-Muñiz, Humberto. "Interregional Transfer of Biological Technology in the Caribbean: The Impact of Barbados' John R. Bovell's Cane Research on the Puerto Rican Sugar Industry, 1888–1920s." *Revista Mexicana del Caribe* 2, no. 3 (1997): 6–40.

Garesche, W. A. *Complete Story of the Martinique and St. Vincent Horrors*. [New York?]: L. G. Stahl, 1902.

Gaspar, David Barry. *Bondmen and Rebels: A Study of Master-Slave Relations in Antigua*. Baltimore: Johns Hopkins University Press, 1985.

Golway, Terry. *So Others Might Live: A History of New York's Bravest, the FDNY from 1700 to the Present*. New York: Basic Books, 2002.

Gordon, Shirley C. *A Century of West Indian Education*. London: Longman, 1963.

Goudsblom, Johan. *Fire and Civilization*. New York: Penguin, 1992.

Goveia, E. V. *Slave Society in the British Leeward Islands at the End of the Eighteenth Century*. 1965. Reprint, Westport, Conn.: Greenwood Press, 1980.

———. *The West Indian Slave Laws of the Eighteenth Century*. Barbados: Caribbean Universities Press, 1970.

Green, William A. *British Slave Emancipation: The Sugar Colonies and the Great Experiment, 1830–1865*. Oxford: Clarendon Press, 1976.

Grenada Handbook, Directory, and Almanac for the Year 1910. London: Wyman and Sons, 1910.

Grieve, Symington. *Notes upon the Island of Dominica*. London: A&C Block, 1906.

Grossman, Lawrence S. *The Political Ecology of Bananas: Contract Farming, Peasants, and Agrarian Change in the Eastern Caribbean*. Chapel Hill: University of North Carolina Press, 1998.

Grove, Richard H. *Green Imperialism: Colonial Expansion, Tropical Island Edens, and the Origins of Environmentalism, 1600–1860*. Cambridge: Cambridge University Press, 1995.

———. "Origins of Western Environmentalism." *Scientific American* 267, no. 1 (1992): 42–46.

Hagelberg, G. B. *The Caribbean Sugar Industries: Constraints and Opportunities*. New Haven: Yale University Antilles Research Program, 1974.

Handler, Jerome S. "Plantation Slave Settlements in Barbados, 1650s to 1834." In *In the Shadow of the Plantation: Caribbean History and Legacy. In Honour of Professor Emeritus Woodville K. Marshall*, edited by Alvin O. Thompson, 121–61. Kingston, Jamaica: Ian Randle, 2002.

Handler, Jerome S., and Frederick W. Lange. *Plantation Slavery in Barbados: An Archaeological and Historical Investigation*. Cambridge: Harvard University Press, 1978.

Hanson, Neil. *The Great Fire of London in That Apocalyptic Year, 1666*. Hoboken, N.J.: John Wiley and Sons, 2002.

Haraksingh, Kusha. "Control and Resistance among Indian Workers." In *India in the Caribbean*, edited by David Dabydeen and Brinsley Samaroo, 61–80. Hansib: University of Warwick Publications, 1987.

———. "Labour, Technology, and the Sugar Estates in Trinidad, 1870–1914." In *Crisis and Change in the International Sugar Economy, 1860–1914*, edited by Bill Albert and Adrian Graves, 133–37. Norwich, England: ISC Press, 1984.

Hardy, Anne. "Water and the Search for Public Health in London in the Eighteenth and Nineteenth Centuries." *Medical History* 28 (1984): 250–82.

Harris, David R. *Plants, Animals, and Man in the Outer Leeward Islands, West Indies: An Ecological Study of Antigua, Barbuda, and Anguilla*. Berkeley: University of California Press, 1965.

Harrison, Lucia Carolyn. "Dominica: A Wet Tropical Human Habitat." *Economic Geography* 11 (1935): 62–76.

Hart, J. H. "Report on the Agri-Horticultural Resources of Tobago." *Bulletin of Miscellaneous Information, Trinidad, Royal Botanic Gardens*, no. 12 (1889).

———. *Trinidad: Report on Forest Conservation*. London: Waterlow and Sons, 1891.

Hazen, Margaret H., and Robert M. Hazen. *Keepers of the Flame: The Role of Fire in American Culture*. Princeton: Princeton University Press, 1992.

Healy, David. *Drive to Hegemony: The United States in the Caribbean, 1898–1917*. Madison: University of Wisconsin Press, 1988.

Herskovits, Melville J., and Frances S. Herskovits. *Trinidad Village*. New York: Knopf, 1947.

Heuman, Gad. *"The Killing Time": The Morant Bay Rebellion in Jamaica*. Knoxville: University of Tennessee Press, 1994.

Higgins, George E. *A History of Trinidad Oil*. [Port of Spain]: Trinidad Express Newspapers Limited, 1996.

Higman, B. W. *Slave Populations of the British Caribbean, 1807–1834*. Baltimore: Johns Hopkins University Press, 1984.

———, ed. *General History of the Caribbean*. Vol. 6. London: UNESCO Publishing, Macmillan, 1999.

Higuera-Gundy, Antonia, Mark Brenner, David A. Hodell, Jason H. Curtis, Barbara W. Leyden, and Michael W. Binford. "A 10,300 C-14 Year Record of Climate and Vegetation Change from Haiti." *Quaternary Research* 52, no. 2 (1999): 159–70.

Hill, Robert T. *Cuba and Porto Rico, with the Other Islands of the West Indies*. London: T. Fisher Unwin, 1898.

Hobsbawm, Eric J. *Labouring Men: Studies in the History of Labour*. Garden City, N.Y.: Anchor, 1967.

Hodge, W. H. "The Vegetation of Dominica." *Geographical Review* 33, no. 3 (1943): 349–75.

Hooper, E. D. M. *Report upon Antigua in Relation to Forestry*. Madras: Lawrence Asylum Press, 1888.

———. *Report upon the Forests of Grenada and Carriacou*. London: Waterlow and Sons, 1887.

———. *Report upon the Forests of Honduras*. Kurnool, India: Collectorate Press, 1887.

———. *Report upon the Forests of Jamaica*. London: Waterlow and Sons, 1886.

———. *Report upon the Forests of St. Lucia*. Madras: Lawrence Asylum Press, 1887.

———. *Report upon the Forests of St. Vincent*. London: Waterlow and Sons, 1886.

———. *Report upon the Forests of Tobago*. Madras: Lawrence Asylum Press, 1887.

Hough, Walter. *Fire as an Agent in Human Culture*. Washington, D.C.: Government Printing Office, 1926.

Howard, Richard A. "The Vegetation of the Antilles." In *Vegetation and Vegetational History of Northern Latin America*, edited by Alan Graham, 1–38. Amsterdam: Elsevier, 1973.

Hulme, Peter. *Colonial Encounters: Europe and the Native Caribbean, 1492–1797*. London: Methuen, 1986.

Innes, F. C. "The Pre-Sugar Era of European Settlement in Barbados." *Journal of Caribbean History* 1 (1970): 1–22.

James, C. L. R. *The Black Jacobins: Toussaint L'Ouverture and the San Domingo Revolution*. 3rd ed. London: Allison and Busby, 1980.

Johnson, Howard. "The Origins and Early Development of Cane Farming in Trinidad." *Journal of Caribbean History* 5 (1972): 46–73.

Joseph, Cedric L. "The Strategic Importance of the British West Indies, 1882–1932." *Journal of Caribbean History* 6 (1973): 23–67.

Karch, Cecelia, with Henderson Carter. *The Rise of the Phoenix: The Barbados Mutual Life Assurance Society in Caribbean Economy and Society, 1840–1990*. Kingston, Jamaica: Ian Randle, 1997.

Kimber, Clarissa Thérèse. *Martinique Revisited: The Changing Plant Geographies of a West Indian Island*. College Station: Texas A&M University Press, 1988.

Kiple, Kenneth F. *The Caribbean Slave: A Biological History*. Cambridge: Cambridge University Press, 1984.

Knight, Franklin W. *The Caribbean: The Genesis of a Fragmented Nationalism*. New York: Oxford University Press, 1978.

Koeppel, Gerard T. *Water for Gotham: A History*. Princeton: Princeton University Press, 2000.

Kuhlken, Robert. "Settin' the Woods on Fire: Rural Incendiarism as Protest." *Geographical Review* 89, no. 3 (1999): 343–63.

Langley, Lester D. *The United States and the Caribbean in the Twentieth Century*. Athens: University of Georgia Press, 1982.

Laurence, K. O. "The Trinidad Water Riot of 1903: Reflections of an Eye Witness." *Caribbean Quarterly* 15, no. 4 (1969): 5–22.

Le Bon, Gustave. *The Crowd: A Study of the Popular Mind*. Introduction by Robert K. Merton. 1895. Reprint, New York: Viking Press, 1960.

Lent, John A. *Mass Communications in the Caribbean*. Ames: Iowa State University Press, 1990.

Levy, Claude. *Emancipation, Sugar, and Federalism: Barbados and the West Indies, 1833–1876*. Gainesville: University Presses of Florida, 1980.

Lowenthal, David. *West Indian Societies*. New York: Oxford University Press, 1972.

MacArthur, Robert H., and Edward O. Wilson. *The Theory of Island Biogeography*. Princeton: Princeton University Press, 1967.

Magid, Alvin. *Urban Nationalism: A Study of Political Development in Trinidad*. Gainesville: University Presses of Florida, 1988.

Manning, Frank E. "Nicknames and Number Plates in the British West Indies." *Journal of American Folklore* 87 (1974): 123–32.

Masefield, G. B. *A History of the Colonial Agricultural Service*. Oxford: Clarendon Press, 1972.

McClelland, J. S. *The Crowd and the Mob: From Plato to Canetti*. London: Unwin Hyman, 1989.

McCullough, David. *The Path between the Seas: The Creation of the Panama Canal, 1870–1914*. New York: Simon and Schuster, 1977.

McDonald, Roderick A. *The Economy and Material Culture of Slaves: Goods and Chattels on the Sugar Plantations of Jamaica and Louisiana*. Baton Rouge: Louisiana State University Press, 1993.

Moreno Fraginals, Manuel. *The Sugarmill: The Socioeconomic Complex of Sugar in Cuba*. Translated by Cedric Belfrage. New York: Monthly Review Press, 1976.

Morison, Samuel Eliot. *Admiral of the Ocean Sea: A Life of Christopher Columbus*. New York: Time, 1962.

Morris, Daniel. *Forest Conservation in Jamaica: Jamaica, Forests, 1877–96*. Bound volume in the archives of the Royal Botanic Gardens, Kew, Surrey, England, 1882.

———. "The Natural History of the Sugar Cane." In *Lectures to Sugar Planters*, edited by the Imperial Commissioner of Agriculture for the West Indies, 1–26. London: Dulau and Co., 1906.

Murray, John A., ed. *The Islands and the Sea: Five Centuries of Nature Writing from the Caribbean*. New York: Oxford University Press, 1991.

Musicant, Ivan. *Empire by Default: The Spanish-American War and the Dawn of the American Century*. New York: Henry Holt, 1998.

Newson, Linda A. *Aboriginal and Spanish Colonial Trinidad: A Study in Culture Contact*. London: Academic Press, 1976.

Nicholls, H. A. Alford. "Legislation to Control Bush Fires." *West Indian Bulletin* 2 (1901): 79–96.

———. *A Text-Book of Tropical Agriculture*. London: Macmillan, 1892.

Oliver, Vere Langford. *The History of the Island of Antigua*. 3 vols. London: Mitchell and Hughes, 1894–99.

Ortiz, Fernando. *Cuban Counterpoint: Tobacco and Sugar*. 1947. Reprint, Durham: Duke University Press, 1995.

Pactor, Howard S., comp. *Colonial British Caribbean Newspapers: A Bibliography and Directory*. New York: Greenwood Press, 1990.

Pande, I. D., and Deepa Pande. "Forestry in India through the Ages." In *History of Forestry in India*, edited by Ajay S. Rawat, 151–62. New Delhi: Indus, 1991.

Pares, Richard. *A West-India Fortune*. 1950. Reprint, London: Archon, 1968.

Pattullo, Polly. *Fire from the Mountain: The Tragedy of Montserrat and the Betrayal of Its People*. London: Constable, 2000.

Pearse, Andrew. "Carnival in Nineteenth Century Trinidad." *Caribbean Quarterly* 4, no. 3 (1955): 175–93.

Pérez, Louis A. *Winds of Change: Hurricanes and the Transformation of Nineteenth-Century Cuba*. Chapel Hill: University of North Carolina Press, 2001.

Petersen, James B. "Taino, Island Carib, and Prehistoric Amerindian Economies in the West Indies: Tropical Forest Adaptations to Island Environments." In *The Indigenous People of the Caribbean*, edited by Samuel M. Wilson, 118–30. Gainesville: University Press of Florida, 1997.

Pitman, Frank Wesley. "Slavery on British West India Plantations in the Eighteenth Century." *Journal of Negro History* 11 (1926): 584–668.

Prestoe, Henry. *Report on Coffee Cultivation in Dominica*. Trinidad: Government Printery Office, 1875.

Pullen-Burry, B. *Ethiopia in Exile: Jamaica Revisited*. London: T. Fisher Unwin, 1905.

Pyne, Stephen J. *Burning Bush: A Fire History of Australia*. New York: Holt, 1991.

——. *Fire in America: A Cultural History of Wildland and Rural Fire*. Princeton: Princeton University Press, 1982.

——. *Vestal Fire: An Environmental History, Told through Fire, of Europe and Europe's Encounter with the World*. Seattle: University of Washington Press, 1997.

——. *World Fire: The Culture of Fire on Earth*. New York: Holt, 1995.

Quilter, Sir Cuthbert. *Notes on a Visit to Some of the West Indian Islands*. London: Waterlow and Sons, 1899.

Ragatz, Lowell J. *The Fall of the Planter Class in the British Caribbean, 1763–1833: A Study in Social and Economic History*. New York: Century, 1928.

Ramdin, Ron. *Arising from Bondage: A History of the Indo-Caribbean People*. New York: New York University Press, 2000.

Ramesar, Marianne. "Patterns of Regional Settlement and Economic Activity by Immigrant Groups in Trinidad, 1851–1900." *Social and Economic Studies* 25, no. 3 (1976): 187–215.

Raymond, Nathaniel. "Cane Fires on a British West Indian Island." *Social and Economic Studies* 16, no. 3 (1967): 280–88.

Report of the West India Royal Commission. London: HMSO, 1897.

Reports on the Botanic Station, Agricultural Instruction, and Experiment Plots, Grenada, 1906–1907. Barbados: Imperial Commission of Agriculture for the West Indies, 1907.

Ribbentrop, B. *Forestry in British India*. Calcutta: Office of the Superintendent of Government Printing, India, 1900.

Richards, Glen. "Collective Violence in Plantation Societies: The Case of the St. Kitts Labour Protests of 1896 and 1935." Institute of Commonwealth Studies, University of London, 1987. Mimeographed.

——. "Masters and Servants: The Growth of the Labour Movement in St. Christopher–Nevis, 1896 to 1956." Ph.D. diss., University of Cambridge, 1989.

——. "Order and Disorder in Colonial St. Kitts: The Role of the Armed Forces in Maintaining Labour Discipline, 1896–1935." Paper presented at the 25th Annual Conference of the Association of Caribbean Historians, Mona, Jamaica, March 27–April 2, 1993.

Richardson, Bonham C. *Caribbean Migrants: Environment and Human Survival on St. Kitts and Nevis*. Knoxville: University of Tennessee Press, 1983.

——. "Catastrophes and Change on St. Vincent." *National Geographic Research* 5, no. 1 (1989): 111–25.

——. "Depression Riots and the Calling of the 1897 West India Royal Commission." *Nieuwe West-Indische Gids* 66, nos. 3 and 4 (1992): 169–91.

——. "Detrimental Determinists: Applied Environmentalism as Bureaucratic Self-Interest in the Fin-de-Siècle British Caribbean." *Annals of the Association of American Geographers* 86, no. 2 (1996): 213–34.

——. *Economy and Environment in the Caribbean: Barbados and the Windwards in the Late 1800s*. Gainesville: University Press of Florida; Kingston, Jamaica: Press University of the West Indies, 1997.

——. "Human Mobility in the Windward Caribbean, 1884–1902." *Plantation Society* 2, no. 3 (1989): 301–19.

——. *Panama Money in Barbados, 1900–1920*. Knoxville: University of Tennessee Press, 1985.

——. "A 'Respectable' Riot: Guy Fawkes Night in St. George's, Grenada, 1885." *Journal of Caribbean History* 27 (1993): 21–35.

Roberts, Peter A. *West Indians and Their Language*. Cambridge: Cambridge University Press, 1988.

Rose, Jonathan. *The Intellectual Life of the British Working Classes*. New Haven: Yale University Press, 2001.

Rosen, Christine Meisner. *The Limits of Power: Great Fires and the Process of City Growth in America*. Cambridge: Cambridge University Press, 1986.

Rossotti, Hazel. *Fire*. Oxford: Oxford University Press, 1993.

Rouse, Irving. *The Tainos: Rise and Decline of the People Who Greeted Columbus*. New Haven: Yale University Press, 1992.

Rudé, George F. E. *The Crowd in History: A Study of Popular Disturbances in France and England, 1730–1848*. New York: John Wiley, 1964.

Salisbury, Harrison E. *The Great Black Dragon Fire: A Chinese Inferno*. Boston: Little, Brown, 1989.

Sauer, Carl Ortwin. *The Early Spanish Main*. Berkeley: University of California Press, 1966.

——. "Man's Dominance by Use of Fire." *Geoscience and Man* 10 (1975): 1–13.

Schwartz, Stuart B. "The Hurricane of San Ciriaco: Disaster, Politics, and Society in Puerto Rico, 1899–1901." *Hispanic American Historical Review* 72, no. 3 (1992): 303–34.

Scott, Stephen Kingsley. "Through the Diameter of Respectability: The Politics of Historical Representation in Postemancipation Colonial Trinidad." *Nieuwe West-Indische Gids* 76, nos. 3 and 4 (2002): 271–304.

Shephard, C. Y. "The Sugar Industry of the British West Indies and British Guiana with Special Reference to Trinidad." *Economic Geography* 5 (1929): 149–75.

Sheridan, Richard B. *Doctors and Slaves: A Medical and Demographic History of Slavery in the British West Indies, 1680–1834*. Cambridge: Cambridge University Press, 1985.

——. *Sugar and Slavery: An Economic History of the British West Indies, 1623–1775*. Baltimore: Johns Hopkins University Press, 1974.

Simpson, George Eaton. "The Nine-Night Ceremony in Jamaica." *Journal of American Folklore* 70 (1957): 329–35.

Smith, Keithlyn B., and Fernando C. Smith. *To Shoot Hard Labour: The Life and Times of Samuel Smith, an Antiguan Workingman, 1877–1982*. Scarborough, Canada: Edan's Publishers, 1986.

Smorkaloff, P. M. *If I Could Write This in Fire: An Anthology of Literature from the Caribbean*. New York: New Press, 1994.

Southey, Thomas. *Chronological History of the West Indies*. 3 vols. London: Longman, Rees, Orme, Brown, and Green, 1827.

Stewart, J. *A View of the Past and Present State of the Island of Jamaica*. 1823. Reprint, New York: Negro Universities Press, 1969.

Stubbs, William C. *Sugar Cane: A Treatise on the History, Botany, and Agriculture of Sugar Cane*. N.p.: Louisiana State Bureau of Agriculture and Immigration, 1897.

Thomas, Hugh. *The Slave Trade: The Story of the Atlantic Slave Trade, 1440–1870*. New York: Simon and Schuster, 1997.

Thomas, J. J. *Froudacity: West Indian Fables by James Anthony Froude*. 1889. Reprint, London: New Beacon Books, 1969.

Tilchin, William N. *Theodore Roosevelt and the British Empire: A Study in Presidential Statecraft*. New York: St. Martin's Press, 1997.

———. "Theodore Roosevelt, Anglo-American Relations, and the Jamaica Incident of 1907." *Diplomatic History* 19, no. 3 (1995): 385–405.

Traxel, David. *1898: The Birth of the American Century*. New York: Knopf, 1998.

Trotman, David Vincent. *Crime in Trinidad: Conflict and Control in a Plantation Society, 1838–1900*. Knoxville: University of Tennessee Press, 1986.

Uncle Sam's Navy. Philadelphia: Historical Publishing Co., 1898.

Vale, Thomas R., ed. *Fire, Native Peoples, and the Natural Landscape*. Washington, D.C.: Island Press, 2002.

Walker, D. J. R. *Columbus and the Golden World of the Island Arawaks: The Story of the First Americans and Their Caribbean Environment*. Sussex: Book Guild, 1992.

Watts, David. *The West Indies: Patterns of Development, Culture, and Environmental Change since 1492*. Cambridge: Cambridge University Press, 1987.

Watts, Francis. "Tree Planting in Antigua." *West Indian Bulletin* 1 (1900): 402–14.

"The West Indies: A Warning and a Way." *International Sugar Journal* 4, no. 46 (1902): 522–25.

White, Timothy. *Catch a Fire: The Life of Bob Marley*. New York: Holt, Rinehart and Winston, 1983.

Wiles, Robert. *Cuban Cane Sugar*. Indianapolis: Bobbs-Merrill, 1916.

Wilkinson, Alec. *Big Sugar: Seasons in the Cane Fields of Florida*. New York: Vintage, 1990.

Will, H. A. "Colonial Policy and Economic Development in the British West Indies, 1895–1903." *Economic History Review* 23 (1970): 129–47.

Williams, Eric. *History of the People of Trinidad and Tobago*. Trinidad: PNM, 1962.

Williams, Joseph J. *Voodoos and Obeahs: Phases of West India Witchcraft*. London: George Allen and Unwin, 1933.

Williams, Sir Ralph. *How I Became a Governor*. London: John Murray, 1913.

Wimbush, A. *Report on the Forestry Problems of the Windward and Leeward Islands*. Trinidad: Government Printer, 1936.

Yates, Ann Watson. *Bygone Barbados*. Barbados: Blackbird Studios, 1998.

Yergin, Daniel. *The Prize: The Epic Quest for Oil, Money, and Power*. New York: Simon and Schuster, 1993.

Zebrowski, Ernest. *The Last Days of St. Pierre: The Volcanic Disaster That Claimed Thirty Thousand Lives*. New Brunswick, N.J.: Rutgers University Press, 2002.

Index